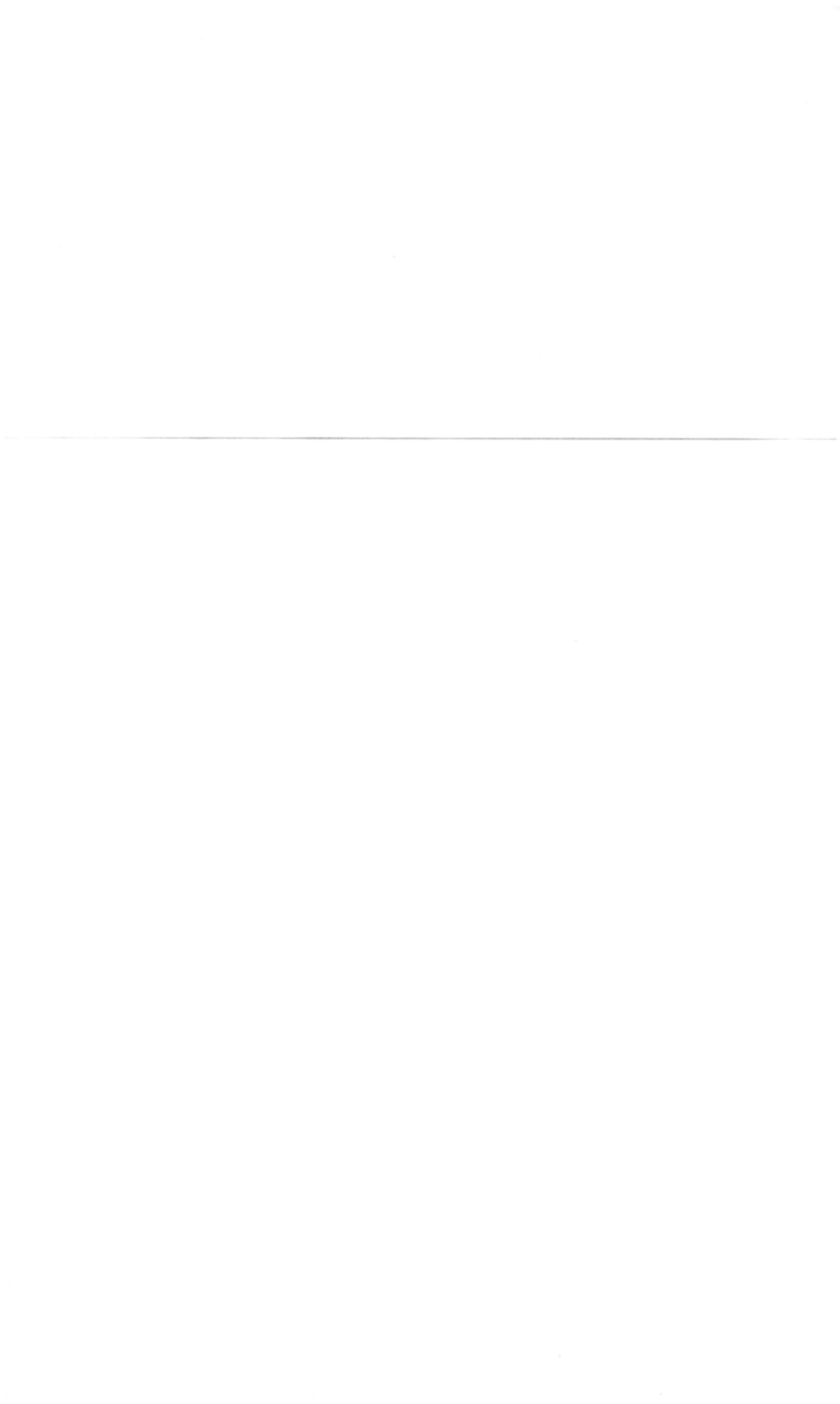

Signal Processing and Image Processing for Acoustical Imaging

Woon Siong Gan

Signal Processing and Image Processing for Acoustical Imaging

Woon Siong Gan
Acoustical Technologies Singapore Pte Ltd
Singapore, Singapore

ISBN 978-981-10-5549-2 ISBN 978-981-10-5550-8 (eBook)
https://doi.org/10.1007/978-981-10-5550-8

This Springer imprint is published by the registered company Springer Nature Singapore Pte Ltd.
The registered company address is: 152 Beach Road, #21-01/04 Gateway East, Singapore 189721, Singapore

To My Family

Preface

Acoustical imaging has been growing rapidly during the last 40 years. It is nowadays a very useful tool for non-destructive testing and medical ultrasound imaging. Its status is catching up fast with X-ray and is at least now on par with X-ray on the market share as it has the advantage of causing no radiation damage. The purpose of writing this book is to be a supplement to my book *Acoustical Imaging: Techniques & Applications for Engineers* published by John Wiley & Sons in 2012. Signal processing and image processing are essential tools in acoustical imaging as they help to sharpen the images and improve the image resolution. So far, there are several books already available in the market on these subjects but they are mostly on the fundamentals and the general applications without a book solely tailored, for acoustical imaging. During the process of writing this book, care has been taken to avoid excessive mathematical equations and to be more readable to the general public. The book will be used as a reference book for practising engineers in the field of acoustical imaging as well as a textbook for advanced undergraduates and first-year graduate students.

The topics on signal processing and image processing in this book are popular and basic in nature without plunging into advanced topics which are heavily mathematically involved. Being a short book, it will be more economical to purchase as well as enabling faster coverage.

With the arrival of artificial intelligence, there is a chapter in this book on the application of deep learning and neural network to acoustical imaging and also there is a chapter on examples of practical application of to signal processing and image processing.

Singapore — Woon Siong Gan

Contents

Chapter 1
What is Signal?

1.1 Introduction

The purpose of this book is on signal processing and image processing with applications to acoustical imaging. The purpose is the use of techniques and methods in signal processing and in image processing to filter out the unwanted noise and to enhance the quality of acoustical images.

Signal carries information. Signal has a broad coverage and can include image, speech, audio, sonar, medical, video, radar, musical, geophysical, and communication. It enables the communication, processing, analysis, and representation of information. It carries the information on the behaviour and attributes of some phenomena. It is an electronic function. A very common example is in electronics and in telecommunication, signal will be on the form of voltage and current which carry information. Thus the study of the contents in signal is known as communication theory. It can be in the form of electromagnetic wave or acoustic wave.

Signal can be define as any quantity that exhibits variation in time or space and provides information on the status of a physical system and it can convey message between observers.

1.2 Forms of Signals

Depending on its function in the time domain, signals can be classified into continuous time signals and discrete time signals. In turn, continuous time signals are also known as analog signals and discrete time signals are also known as digital time signals. A digital signal is an approximation of an analog signal with values at particular times. Hence, they are quantized. The word analog in analog signal means that the time varying feature of the signal is analogous to another time varying signal it represents. For instance, an analog signal represents the acoustic pressure of the sound wave has a time varying feature analogous to the instantaneous voltage of the audio

W. S. Gan, *Signal Processing and Image Processing for Acoustical Imaging*,
https://doi.org/10.1007/978-981-10-5550-8_1

signal. A digital signal is constructed from a discrete set of waveforms of a physical quantity. It represents a sequence of discrete values. An example of digital signal is logic signal with only two possible values. One can obtain digitised signals through sampling of analog signals. For instance, an analog to digital converter circuit can digitise a continually fluctuating voltage on line.

1.3 Information Theory

Signal carries information and so it is necessary to understand what is information theory. Information is a guided principle for signal processing. Information theory can give rise to advanced methods in signal processing. Signals represent a physical representation of information such that the intended user can access this information with ease. Also signal processing helps to remove irrelevant information and presenting the maximum relevant information. Shannon [1, 2] founded the concept of entropy from statistical founded the information theory by using a mathematical and statistical approach. The concept of entropy from statistical mechanics is used to measure information. He quantified information by using entropy. The concept of entropy plays a central role in information theory and is a key measure in information theory. Entropy quantifies the amount of uncertainty involved in the outcome of a random process. In statistical mechanics, entropy is defined as:

$$\text{Entropy} = \mathrm{S} = k_B \ln \Omega \tag{1.1}$$

where k_B = Boltzmann's constant and Ω = number of microstates.

In statistical mechanics, entropy is an extensive property of a thermodynamic system.

References

. Shannon, C. E. (1948). A mathematical theory of communication. *Bell System Technical Journal*, *27*(3), 379–423.
. Shannon, C. E. (1948). A mathematical theory of communication. *Bell System Technical Journal*, *27*(4), 623–666.

Chapter 2
Fourier Series

2.1 Fourier Analysis

The Fourier series [1] is a time series of representation of sinuisoidal signals. It is linear in nature and hence is a form of superposition. It is composed of cosine series and sine series represented as follows:

$$\begin{aligned} \mathrm{f(t)} &= \cos \omega \mathrm{t} + \cos 2\omega \mathrm{t} + \cos 2(n-1)\omega \mathrm{t} + \ldots \sin \omega \mathrm{t} + \sin 2\omega \mathrm{t} \\ &\quad + \ldots \sin 2(n-1)\omega \mathrm{t} \end{aligned} \tag{2.1}$$

This is a representation in the time domain.

Fourier analysis is the mathematical process of decomposing a mathematical function into sinuisoidal or oscillatory components. Here a broad spectrum of mathematics is involved. It is widely used in science and engineering. The reverse process is known as Fourier synthesis. The process of Fourier analysis is termed Fourier Transform which is branch of study by itself and in this book a separate chapter is devoted to this. Fourier analysis can be extended to include more general situations to form the field of harmonic analysis.

Fourier analysis has many scientific applications arising from the properties of the Fourier Transform. It has been applied to signal processing digital image processing, partial differential equations, statistics, sonar etc. The examples of Fourier analysis in signal processing are in audio signals, seismic wave signals processing and in image processing. In audio signal processing it can be used in the equalization of audio recordings with a series bandpass filter. The generation of sound spectrogram can be used to analyze sounds. In sonar, passive sonar can be used to classify targets based on machinery noise. Acoustical image processing can be used to remove periodic or anisotropic artifacts and the cross-correlation of similar images for co-alignment.

W. S. Gan, *Signal Processing and Image Processing for Acoustical Imaging*,
https://doi.org/10.1007/978-981-10-5550-8_2

2.2 Harmonic Analysis

Harmonic analysis is the study of the Fourier series related to computation making use of the orthogonality relationship of the sine and cosine functions. In this case, it is an expansion of a periodic function of f(x) in terms of an infinite sum of sines and cosines. Hence harmonic is an useful method to break up an arbitrary periodic function into a set of simple terms that be solved individually and then recombined to obtain the solution to the original problem. It can be an approximation to the solution to whatever accuracy is predicted or desired. Then successive approximations can be performed to the periodic function.

2.3 Properties of the Fourier Series

The superposition principle holds for the Fourier series and it holds for solution of a linear homogeneous ordinary differential equation. If the solution of equation is obtained for a single sinuisoid then for an arbitrary function the solution can be obtained by expressing the arbitrary function as a Fourier series and then obtain the solution for each sinuisoidal component. One can represent any periodic signal as a Fourier series of series of sines and cosines. The sines and cosines are of different frequencies. There are two forms of the Fourier series in the frequencies regime: the trigonometric series and the exponential series. Actually these two forms are equivalent

Also the Fourier series has symmetry properties. In the trigonometric series there are the concepts of even functions and odd functions. An even function, x_e(t) is symmetric about t = 0, that is

$$x_e(\mathrm{t}) = x_e(-\mathrm{t}) \tag{2.2}$$

Also an even function x_e(t) can be expressed a sum of cosines of various frequencies:

$$x_e(\mathrm{t}) = \sum_{n=0}^{\infty} a_n \cos(\mathrm{n}\omega\mathrm{t}) \tag{2.3}$$

Since cos(nωt) when n = 0 is 1 so (2.3) can be expressed as

$$x_e(\mathrm{t}) = a_0 + \sum_{n=1}^{\infty} a_n \cos(\mathrm{n}\omega\mathrm{t}) \tag{2.4}$$

An odd function is antisymmetric about t = 0, that is

$$x_0(\mathrm{t}) = -x_0(\mathrm{t}) \tag{2.4}$$

An odd function can be expressed as

$$x_0(\mathrm{t}) = \sum_{n=1}^{\infty} b_n \sin(\mathrm{n}\boldsymbol{\omega}\mathrm{t}) \tag{2.5}$$

2.4 Fourier Coefficients

The Fourier coefficients are a_n and b_n. They can be determined as follows.

To derive a_n, one will start with the synthesis equation of the Fourier series given by

$$x_e(\mathrm{t}) = \sum_{n=0}^{\infty} a_n \cos(\mathrm{n}\boldsymbol{\omega}\mathrm{t})$$

Here $x_e(\mathrm{t})$ is an even function.

Multiply both sides by cos(m$\boldsymbol{\omega}$t),

$$x_e(\mathrm{t})\cos(\mathrm{m}\boldsymbol{\omega}\mathrm{t}) = \sum_{n=0}^{\infty} a_n \cos(\mathrm{n}\boldsymbol{\omega}\mathrm{t})\cos(\mathrm{m}\boldsymbol{\omega}\mathrm{t})$$

Next integrate over one period, then

$$\begin{aligned}\int_T x_e(\mathrm{t})\cos(\mathrm{m}\omega\mathrm{t})\mathrm{dt} &= \int_T \sum_0^{\infty} a_n \cos(\mathrm{n}\boldsymbol{\omega}\mathrm{t})\cos(\mathrm{m}\boldsymbol{\omega}\mathrm{t})\mathrm{dt} \\ &= \sum_{n=0}^{\infty} a_n \int_T \frac{1}{2}[\cos(\mathrm{m}+\mathrm{n})\boldsymbol{\omega}\mathrm{t} + \cos(\mathrm{m}-\mathrm{n})\boldsymbol{\omega}\mathrm{t}]\mathrm{dt} \\ &= \frac{1}{2}\sum_{n=0}^{\infty} a_n \int_T [\cos(\mathrm{m}+\mathrm{n})\mathrm{t} + \cos(\mathrm{m}-\mathrm{n})\boldsymbol{\omega}\mathrm{t}]\mathrm{dt} \\ &= \frac{1}{2}\sum_{n=0}^{\infty} a_n \int_T \cos[(\mathrm{m}-\mathrm{n})\omega\mathrm{t}]\mathrm{dt} \end{aligned} \tag{2.6}$$

Cosine is an even function. This means cos[(m − n)$\boldsymbol{\omega}$t] = cos[(n − m)$\boldsymbol{\omega}$t]. It is time varying and has exactly/m − n/complete oscillations in the interval of integration for $m \neq n$. When $m = n$, $\cos[(m - n)\omega t] = \cos(0) = 1$.

So $\int_T \cos[(m-n)\omega t] = 0$ when $m \neq n$

$$\int_T \cos[(m-n)\omega t] = \int_T 1dt = T \text{ when } m = n.$$

In (2.6) as n goes from 0 to ∞, every term in the summation will be zero except when $m = n$ and the integral equals T. Hence the summation in (2.6) will reduce to $\int_T x_e(t)\cos(m\omega t)dt = \frac{1}{2}a_m t$. So

$$a_m = \frac{2}{T}\int_T x_e(t)\cos(m\omega t)dt \tag{2.7}$$

Using similar procedure,

$$a_n = \frac{2}{T}\int_T x_e(t)\cos(n\omega t)dt \tag{2.8}$$

For the special case of $m = 0$,

Since $\int_T x_e(t)\cos(m\omega t)dt = \int_T \sum_0^{\infty} a_n \cos(n\omega t)\cos(m\omega t)dt$

Substituting in $m = 0$, $\int_T x_e(t)\cos(0 \cdot \omega t)dt = \int_T \sum_0^{\infty} a_n \cos(n\omega t)\cos(0 \cdot \omega t)dt$

With $\cos(0 \cdot \omega t) = 1$, $\int_T x_e(t)dt = \sum_{n=0}^{\infty} a_n \int_T \cos(n\omega t)dt = Ta_0$ or

$$a_0 = \frac{1}{T}\int_T x_e(t)dt \tag{2.9}$$

Hence a_0 is the average value of the function $x_e(t)$.

Next the odd function $x_o(t)$ will be considered.

$$x_o(t) = \sum_{n=1}^{\infty} b_n \sin(n\omega t) \tag{2.10}$$

There is no b_0 term because the average value of an odd function over one period is always zero. Following the same procedure as above, it can be shown that

$$b_n = \frac{2}{T}\int_T x_o(\mathrm{t})\sin(n\omega t)\mathrm{dt} \tag{2.11}$$

2.5 The Trigonometric Series and the Exponential Series are Equivalent

The Fourier series synthesis equation is given by

$$x_T(\mathrm{t}) = a_0 + \sum_{n=1}^{\infty}[a_n\cos(\mathrm{n}\omega\mathrm{t}) + b_n\sin(\mathrm{n}\omega\mathrm{t})] \tag{2.12}$$

The Fourier series analysis equations are given by the Fourier coefficients as follows:

$$a_0 = \frac{1}{T}\int_T x_T(\mathrm{t})\mathrm{dt} = \text{average} \tag{2.13}$$

$$a_n = \frac{2}{T}\int_T x_T(\mathrm{t})\cos(\mathrm{n}\omega\mathrm{t})\mathrm{dt},\quad \mathrm{n} \neq 0 \tag{2.14}$$

$$b_n = \frac{2}{T}\int_T x_T(\mathrm{t})\sin(\mathrm{n}\omega\mathrm{t})\mathrm{dt}, \tag{2.15}$$

The exponential series is a compact representation of the Fourier series using complex exponentials. With the same derivation procedure as that for the Fourier cosine series given above, one can obtain the following synthesis and analysis equations:

$$x_T(\mathrm{t}) = \sum_{n=-\infty}^{+\infty} c_n\exp(j\omega t)\quad \text{Synthesis} \tag{2.16}$$

$$c_n = \frac{1}{T}\int_T x(t)\exp(-jn\omega t)\mathrm{dt}\quad \text{Analysis} \tag{2.17}$$

The advantage of the exponential series is that only a single analysis equation is required compared to the three equations for the trigonometric form.

The following will show that the trigonometric series and the exponential series are equivalent. The trigonometric series are given by (2.12) and (2.16) respectively. If they are equivalent, then

$$a_0 + \sum_{n=1}^{\infty}[a_n \cos(n\omega t) + b_n \sin(n\omega t)] = \sum_{n=-\infty}^{+\infty} c_n \exp(jn\omega t)$$

Starting with the constant term, $c_0 = a_0 =$ averaged of the function x_T. Considering only the parts of the signal that oscillate only once in a period of T seconds or n = 1, then

$$a_1 \cos(\omega t) + b_1 \sin(\omega t) = c_{-1} \exp(-j\omega t) + c_1 \exp(j\omega t)$$

For two oscillations in T seconds, or n = 2, then

$$a_2 \cos(2\omega t) + b_2 \sin(2\omega t) = c_{-2} \exp(-j2\omega t) + c_2 \exp(2\omega t)$$

In general,

$$a_n \cos(n\omega t) + b_n \sin(n\omega t) = c_{-n} \exp(-jn\omega t) + c_n \exp(jn\omega t) \tag{2.18}$$

From (2.18), writing $c_n = c_{n,r} + j \cdot c_{n,i}$, then $c_{-n} =$ complex conjugate of $c_n = c * n = c_{n,r} - jc_{n,i}$, then

$$\begin{aligned}
&(c_{n,r} - jc_{n,i})\left[\cos(n\omega t)\, j \sin(n\omega t)\right] + (c_{n,r} + jc_{n,i})[\cos(n\omega t) + j \sin(n\omega t)\\
&= 2c_{n,r} \cos(n\omega t) - 2c_{n,i} \sin(n\omega t) + j \cos(n\omega t)(c_{n,r} - c_{n,i}) + \sin(n\omega t)(c_{n,i} - c_{n,r})\\
&= 2c_{n,r} \cos(n\omega t) - 2c_{n,i} \sin(n\omega t)
\end{aligned}$$

Equating the magnitude of cosine and sine terms with $n \neq 0$, one obtains

$$a_n = 2c_{n,r} \text{ and } b_n = -2c_{n,i}$$

Then $c_n = a_n/2 - jb_n/2$ for $n \neq 0$ with $c_{-n} = c_n^*$.

There are limitations in the applications of Fourier series and one requires them to be convergent with the following conditions to be fulfilled:

1. x_T to have a finite number of discontinuities in one period
2. x_T to have a finite number of maxima and minima in one period
3. $\int_T x_T(t)dt \angle\infty$, which means the function is not infinite over a finite interval.

Reference

1. Wikipedia. Fourier Series.

Chapter 3
Fourier Transform

3.1 The Representation of the Fourier Transform

Fourier Transform [1, 2] decomposes a signal which is a function of time into its constituent frequencies. Fourier Transform refers to the frequency domain representation. It is a mathematical operation, in a function of time. This produces a complex-valued function of frequency. The magnitude or modulus of this complex function is the amount of the frequency present in the original function. The argument of this complex function is the phase offset of the basic sinusoid in the frequency. The domain of the original function of time is the time domain. There is also the inverse Fourier Transform which represents an inverse process of synthesizing the original function of time from its frequency domain representation.

The study of Fourier Transform originates from the study of Fourier series where complicated periodic functions are expressed as sum of sine series and cosine series meaning complex period waves can be represented as superposition of sine waves and cosine waves. This can be treated as an extension of the Fourier series when the period of the represented function is lengthened to approach infinity. The Fourier Transform is expressed as an integral and it enables to recover the amplitude of each wave due to the properties of sine and cosine. One uses the Euler's formula $e^{i2\pi\theta} = \cos(2\pi\theta) + i\ \sin(2\pi\theta)$ to express the Fourier series in terms of the basic waves $e^{i2\pi\theta}$. This will simplify many of the formulae involved. It also makes it necessary for the Fourier coefficients to be complex values. The complex number will provide both the amplitude or size of the waves present in the function and the phase of the waves.

For functions that are zero outside an interval, there is a connection between the definition of Fourier Transform and Fourier series. Then one can calculate its Fourier series on any interval that includes the points where f is not identically zero. When the length of the interval in which one calculates the Fourier series is increased, then the Fourier series coefficients begin to resemble the Fourier transform and the sum of the Fourier series of f begins to resemble the inverse Fourier transform.

W. S. Gan, *Signal Processing and Image Processing for Acoustical Imaging*,
https://doi.org/10.1007/978-981-10-5550-8_3

Let [−T/2, T/2] be the period containing the interval in which f is not identity zero. Then the nth series coefficient c_n is

$$c_n = \int_{-T/2}^{T/2} f(x)\exp(-2\pi\,\mathrm{jnx/T}) \tag{3.1}$$

The Fourier coefficients are equal to the values of the Fourier transform sampled on a grid of width, 1/T, multiplied by the gird width 1/T.

Thus f can be written as

$$\mathrm{f(x)} = \sum_{-\infty}^{\infty} c_n \exp(2\pi\,\mathrm{jnx/T}) \tag{3.2}$$

Fourier Transform is useful in signal processing because signals are usually expressed in term of time and Fourier Transform is useful for the spectral analysis of time series. The exception is for the case of statistical signal processing, sometimes, Fourier transformation is not applied to the signal itself. For transient signals, it is also advisable to model the signal by a function which is stationary in the sense that its characteristic properties are constant over all time.

3.2 The Definition of Fourier Transform

In the usual sense, the Fourier Transform is the transformation of a mathematical function from time function x(t) to a frequency function X(**ω**). It can be taken as a special case of the Fourier series when T → ∞. One starts with the Fourier series analysis equation:

$$\mathrm{x(t)} = \sum_{-\infty}^{+\infty} c_n \exp(\mathrm{jn}\,\omega_0\mathrm{t}) \tag{3.3}$$

c_n is given by the Fourier series analysis equation

$$c_n = \frac{1}{T}\int_T x(t)\exp(-\mathrm{jn}\,\omega_0\mathrm{t}) \tag{3.4}$$

With T → ∞, ω_0 = fundamental frequency = 2π/t becomes extremely small, and one can replace $\boldsymbol{\omega}$ by $n\omega_0$. Letting $X(\omega) = T\,\mathrm{c_n}$, one has

$$X(\omega) = \int_{-\infty}^{+\infty} x(t)\exp(-j\omega t)dt \tag{3.5}$$

This is the forward Fourier Transform. Likewise, the inverse Fourier Transform can be derive from the Fourier series synthesis equation. As $T \to \infty$, 1/t becomes very large, and one can replace ω by $n\omega_0$ and write $X(\omega) = T\,c_n$. This will produce the inverse Fourier Transform as

$$x(t) = \sum_{-\infty}^{+\infty} X(\omega)\exp(j\omega t)d\omega/2\pi = \frac{1}{2\pi}\int_{-\infty}^{+\infty} X(\omega)\exp(j\omega t)d\omega \tag{3.6}$$

References

1. Wikipedia, Fourier Transform.
2. Bracewell, R. (1965). *The Fourier Transform and it applications*. New York: McGraw-Hill.

Chapter 4
Discrete Fourier Transform

The ordinary Fourier Transform is for a continuous function. The continuous Fourier Transform is difficult to use in real time because in real time, one is dealing with discrete data sampled using some kind of sensors. For instance, the time series from weather, traffic, stocks etc., one is getting the discrete values at each time point (e.g. 1-s, 2-s, 3-s etc.) and one does not know what is in between those samples. One just needs to infer. That is why it is better to have a high sampling frequency of say 100 samples per second instead of 1 sample per second so as to have a more accurate approximation of the real data.

The continuous Fourier Transform is defined as

$$X(f) = \int_{-\infty}^{+\infty} \mathrm{x(t)} \exp(-j2\pi f t)dt \tag{4.1}$$

Extending this to the case of a discrete function, $f(t) \rightarrow f(t_k)$ by letting $f_k \equiv f(t_k)$, where $t_k \equiv k\Delta$, with $k = 0, \ldots \mathrm{N} - 1$, one will obtain the discrete Fourier Transform given by

$$F_n = \sum_{k=0}^{N-1} f_k \exp(-\mathrm{j}2\pi\mathrm{nk/N}) \tag{4.2}$$

The inverse discrete Fourier Transform is then given by:

$$f_k = \frac{1}{N} \sum_{n=0}^{N-1} F_n \exp(\mathrm{j}2\pi\mathrm{kn/N}) \tag{4.3}$$

Discrete Fourier Transforms (DFTs) are extremely useful in digital signal processing because they reveal periodicities in input data as well as the relative strengths of any periodic components. The DFT is one of the most powerful tools in digital signal processing which enables one to find the spectrum of a finite duration signal,

W. S. Gan, *Signal Processing and Image Processing for Acoustical Imaging*,
https://doi.org/10.1007/978-981-10-5550-8_4

the other being digital filtering. DFT can be used to determine the frequency content of a time domain signal such as to analyze the spectrum of the output of an LC oscillator to find out how much noise is present in the produced sine wave. This can be arrived at through the problem of spectrum estimation. Besides this, the DFT has several other applications in DSP.

As an illustration, one starts with a continuous time signal and use a finite number of samples to analyze the frequency content of the continuous time signal. The problem of decomposing this discrete sequence boils down to solving a set of linear equations. This is converting x(t) into x(n) which is the equivalent analog continuous time signal of x(t). One can use samples of $Xe^{j\omega}$ to find an approximation of the spectrum of x(n). The idea of sampling $Xe^{j\omega}$ equally spaced frequency points is in fact the basis of DFT. This sampling is taking place in the frequency domain. $Xe^{j\omega}$ is a function of frequency. The purpose is to find a set of sinusoids which can be added together to produce x(n). DFT is based on sampling Eq. 4.1 at equally spatial frequency points. The DTFT of an input sequence x(n) is given by $Xe^{j\omega}$ which is a periodic function in $\boldsymbol{\omega}$ with a period of 2π.

$$\mathrm{X}e^{j\omega} = \sum_{-\infty}^{+\infty} x(n)e^{-jn\omega} \tag{4.4}$$

The inverse of the DTFT is given by

$$\mathrm{x(n)} = \frac{1}{2\pi}\int_{-\pi}^{\pi} Xe^{j\omega}e^{jn\omega}\mathrm{d}\boldsymbol{\omega} \tag{4.5}$$

One can use (4.4) to find the spectrum of a finite duration signal x(n). However, $Xe^{j\omega}$ given by (4.4) is a continuous function of $\boldsymbol{\omega}$. Hence a digital computer cannot directly use (4.4) to analyse x(n). However, one can use samples of $Xe^{j\omega}$ to find an approximation of the spectrum of x(n). The idea of sampling $Xe^{j\omega}$ at equally spaced frequency points is in fact the basis of the second Fourier technique mentioned above, that is the DFT.

The sampling of $xe^{j\omega}$, the DTFT of x(n) in the frequency domain will lead to the replicas of x(n) in the time domain. Sampling a signal in the time domain will lead to the replicas of the original signal in the frequency domain. The calculation of the spectrum of a finite-duration sequence will clarify the inherent periodic behaviour of DFT representation and it is simple and intuitive. Here one only needs to extract a periodic signal out of the N samples of the finite duration sequence. Next, applying the discrete time Fourier series expansion, one can find the frequency domain representation of the periodic signal. The Fourier coefficients obtained are the same as the DFT coefficients except for a scaling factor.

The discrete time Fourier series of this periodic signal is given by

$$a_k = \frac{1}{N}\sum_{n=0}^{N-1} x(n)\exp(-j\frac{2\pi}{N}\text{kn}) \tag{4.6}$$

where N denotes the period of the signal.
The time domain signal can be obtained as follows:

$$\text{x(n)} = \sum_{k=0}^{N-1} a_k \exp(j\frac{2\pi}{N}\text{kn}) \tag{4.7}$$

Multiplying a_k in Eq. (4.6) by N, one will obtain the DFT coefficients X(k):

$$\text{X(k)} = \sum_{n=0}^{N-1} x(n)\exp(-j\frac{2\pi}{N}\text{kn}) \tag{4.8}$$

The inverse DFT will be given by

$$\text{x(n)} = \frac{1}{N}\sum_{k=0}^{N-1} X(k)\exp(j\frac{2\pi}{N}\text{kn}) \tag{4.9}$$

As an example, let us take an L-sample-long sequence, x(n), representing the analog continuous time signal x(t). The purpose here is to find a set of sinusoids which can be added together to produce x(n).

The DFT is based on sampling the DTFT given by (4.4) at equally spaced frequency points. $Xe^{j\omega}$ is a periodic function in $\boldsymbol{\omega}$ with a period of 2π. If one takes N samples in each period of $Xe^{j\omega}$, the spacing between frequency points will be $\frac{2\pi}{N}$. Hence the frequency of the set of sinusoids that one is looking for will be of the form $\frac{2\pi}{N}$xk, where one can select k = 0, 1 , ..., N − 1. Using complex exponentials similar to Eqs. (4.4) and (4.5), the basis function will be $\exp(\text{j}\frac{2\pi}{N}kn)$. One is looking for a weighted sum of these functions which will give us the original signal x(n) which means that

$$\text{X(n)} = \sum_{k=0}^{N-1} X'(k)\exp(\text{j}\frac{2\pi}{N}\text{kn}), n = 0, 1, \ldots \text{L} - 1 \tag{4.10}$$

where X′(k) denotes the weight used for the complex exponential $\exp(\text{j}\frac{2\pi}{N}\text{kn})$.

For a given L and N, the values of complex exponentials are known and having the value of the time-domain signal, one can calculate the coefficients $X'(k)$.

Reference

1. Wikipedia, Discrete Fourier Transform.

Chapter 5
Fast Fourier Transform

5.1 Introduction to Fast Fourier Transform

The fast Fourier Transform (FFT) is an algorithm that increases the computation speed of the DFT of a sequence or its inverse (DFT) by simplifying its complexity. This is because by computing the DFT and IDFT directly from its definition is often too slow to be practical. Fourier analysis converts a signal from its original domain (time or space) into the frequency domain and vice versa. The DFT on the other hand is the decomposition of a sequence of values into components of different frequencies. For long data sets, in the thousands or millions, the reduction in computation time can be enormous. In the presence of round-off error, many FFT algorithms are much more accurate than evaluating the DFT definition directly. There are many FFT algorithms based on complex number arithmetic to number theory and group theory.

As an illustration of the enormous computation time needed in the DFT, let us take the N point DFT equation for a finite duration sequence, x(n) which is given by:

$$\mathrm{X(k)} = \sum_{n=0}^{N-1} x(n)\exp\left(-j\frac{2\pi}{N}\mathrm{kn}\right) \tag{5.1}$$

First one will show how many multiplications and additions are required to calculate the FFT of a sequence using the above equation. For each DFT coefficient X(k), one will calculate N terms, including $\mathrm{x(0)exp}\left(-\mathrm{j}\frac{2\pi}{N}\mathrm{kx0}\right)$, $\mathrm{x(1)exp}\left(-\mathrm{j}\frac{2\pi}{N}\mathrm{kx1}\right)$, $\ldots\mathrm{x(n-1)exp}\left(-\mathrm{j}\frac{2\pi}{N}\mathrm{Kx(N-1)}\right)$, and then calculate the summation. x(n) is a complex number in general and, hence, each DFT output needs approximately N complex multiplications and N − 1 complex

W. S. Gan, *Signal Processing and Image Processing for Acoustical Imaging*,
https://doi.org/10.1007/978-981-10-5550-8_5

additions. A single complex multiplication itself requires four real multiplications and two real additions. This can be simply verified by considering the multiplication of $d_1 = a_1 + jb_1$ by $d_2 = a_2 + jb_2$ which yields

$$d_1 d_2 = (a_1 a_2 - b_1 b_2) + j(b_1 a_2 + a_1 b_2) \tag{5.2}$$

Moreover, a complex addition itself requires two real additions. Therefore, each DFT coefficient requires 4N real multiplications and 2N + 2(N − 1) = 4N − 2 real additions. To have all the DFT coefficients, we have to compute Eq. (5.1) for all N values of k and therefore an N-point DFT analysis requires $4N^2$ real multiplications and N(4N − 2) real additions. The important point is that the number of calculations of an N-point DFT is proportional to N^2 and this can rapidly increase with N.

The modern, generic FFT algorithm is often credited to the paper of Cooley and Tukey published in 1965 [1]. However, the history of fast algorithms for the DFT can be traced as far back to 1805 to Gauss's unpublished work. However, Gauss's work [2] did not analyze the computation time and used other methods to achieve his goal. Between 1805 and 1965, several versions of the FFT were published by various authors. For instance, Frank Yates [3] published his version known as interaction algorithm is 1932 which provided efficient computation of Walsh and Hadamard transforms. It is still used today in statistical design and analysis of experimental data. Danielson and Lanczos [4] published their work for the computation of DFT for x-ray crystallography, where calculation of Fourier transforms presented a formidable bottleneck, in 1942. Many methods had focused on taking advantage of symmetries to reduce the constant factor for $O(N^2)$ computation. Danielson and Lanczos [4] also used periodicity and doubling trick to obtain O(NlogN) to replace $O(N^2)$. The Cooly and Tukey [1] algorithm is applicable when N is composite and not necessarily a power of 2. Tukey came up with this idea during the meeting of President Kennedy's Science Advisory Committee on the discussion of the analyzing of the output of sensors to be set up to detect nuclear tests by the Soviet Union. Here the fast Fourier transform would be needed. Richard Garwin [5] during a discussion with Tukey, that the FFT algorithm is applicable to not just the national security problems, but also to a general applicability including the determination of the periodicity of the spin orientations in a 3 D crystal of Helium. Garwin [5] passed Tukey's idea to Cooley who worked in IBM's Watson labs with him. Cooley and Tukey [1] published the paper within six months. This FFT algorithm was unable to be patented as Tukey did not work in IBM, and it went into the public domain. Through the computing revolution of the 1970s, it became one of the most important algorithms in digital signal processing.

5.2 The Definition of Fast Fourier Transform

The discrete Fourier transform (DFT) is defined as:

$$\mathrm{X}(k) = \sum_{n=0}^{N-1} x(n)\exp(-i2\pi \mathrm{kn/N}),\ \ \mathrm{k} = 0, 1, 2, 3, \ldots, \mathrm{N}-1 \qquad (5.3)$$

and $x_0, \ldots x_{N-1}$ are complex numbers.

To evaluate this definition directly requires $O(N^2)$ operations. This is because there are N outputs x_K and each output requires sum of N terms. The FFT algorithm will enable the computation to obtain the same results by following the O(NlogN) operations instead. This will result in a tremendous reduction in computation time. As an illustration, consider counting the number of complex multiplications and additions for N = 4096 data points. Evaluating the DFT's sums directly involves N^2 complex multiplications and N(N − 1) complex additions of which O(N) operations can be saved by eliminating trivial operations such as multiplications by 1, leaving about 30 millions. On the other hand, using the radix- 2 Cooley-Tukey algorithm, for N a power of 2, can compute the same result with only (N/2)$\log_2$(N) complex multiplications (ignoring multiplications by 1) and N $\log_2$ N complex additions, a total of about only 30,000 operations which is a thousand times less than with direct evaluation. This shows the improvement from $O(N^2)$ to O(N log N) operations.

5.3 Fast Fourier Transform Algorithms

Sofar the most widely used FFT algorithm is the Cooley-Tukey algorithm [1]. Fast Fourier Transform algorithms generally fall into two classes: decimation in time, and decimation in frequency. The Cooley-Tukey FFT algorithm [1] first rearranges the input elements in bit-reversed order, then builds the output transform. This belongs to decimation in time. The basic idea is to break up a transform of length N into two transforms of length N/2 using the identity

$$\begin{aligned}\sum_{n=0}^{N-1} a_n e^{-2\pi ink/N} &= \sum_{n=0}^{N/2-1} a_{2n} e^{-2\pi i(2n)k/N} \\ &\quad + \sum_{n=0}^{N/2-1} a_{2n-1} e^{-2\pi i(2n+1)k/N} \\ &= \sum_{n=0}^{N/2-1} a_n^{\mathrm{even}} e^{-2\pi ink/(N/2)}\end{aligned}$$

$$+\,e^{-2\pi ik/N}\sum_{n=0}^{N/2-1} a_n^{\mathrm{odd}} e^{-2\pi ink/(N/2)}, \tag{5.4}$$

This is known as the Danielson-Lanczos lemma.

A discrete Fourier Transform can be computed using an FFT by means of the Danielson-Lanczos lemma [3] if the number of points N is a power of two. If the number of points N is not a power of two, a transform can be performed on sets of points corresponding to the prime factors of N which is slightly degraded in speed. An efficient real Fourier Transform algorithm gives a further increase in speed by approximately a factor of two. Base-4 and base-8 fast Fourier Transform use optimized code, and can be 20–30% faster than base-2 fast Fourier Transform.

References

1. Cooley, J. W., & Tukey, J. W. (1965). An algorithm for the machine calculation of complex fourier series. *Mathematics of Computation, 19*(90), 297–301.
2. Gauss, C. F. (1866). Theoria interpolationis methodo nova tractata [Theory regarding a new method of interpolation]. *Nachlass* (Unpublished manuscript). Werke (in Latin and German). 3. Göttingen, Germany: Königlichen Gesellschaft der Wissenschaften zu Göttingen. pp. 265–303.
3. Yates, F. (1937). The design and analysis of factorial experiments. Technical Communication no. 35 of the Commonwealth Bureau of Soils.
4. Danielson, G. C., & Lanczos, C. (1942). Some improvements in practical fourier analysis and their application to x-ray scattering from liquids. *Journal of the Franklin Institute, 233*(4), 365–380.
5. Garwin, R. (1969) The fast fourier transform as an example of the difficulty in gaining wide use for a new technique. *IEEE Transactions on Audio and Electroacoustics, AU-17*(2), 68.

Chapter 6
Convolution, Correlation, and Power Spectral Density

6.1 Introduction to Convolution

Convolution [1] is the single most important technique in digital signal processing (DSP). It provides the mathematical framework for DSP. It is a mathematical method of combining two signals to form a third signal. It related the three signals of interest: the input signal, the output signal, and the impulse response. Using the strategy of impulse decomposition, systems are described by a signal called the impulse response. Convolution is a mathematical operation on two functions f and g to produce a third function that expresses how the shape of one is modified by the other. It refers to both the result function and to the procedure of computing it. It is defined as the integral of the product of the two functions after one is reversed and shifted. Convolution has been applied to signal processing, image processing, differential equations, statistics, etc.

6.2 Definition of Convolution

The convolution of f and g is written $f * g$, using an asterisk. It is defined as the integral of the product of the two functions f and g after one is reversed and shifted. As such, it is a particular kind of integral transform:

$$(\mathrm{f} * \mathrm{g}) = \int_{-\infty}^{\infty} f(\tau)\mathrm{g}(\mathrm{t}-\tau)\mathrm{d}\tau \tag{6.1}$$

W. S. Gan, *Signal Processing and Image Processing for Acoustical Imaging*,
https://doi.org/10.1007/978-981-10-5550-8_6

As an illustration, one will show that the convolution of two Gaussian functions:

$$f = e^{-(t-\mu_1)^2/(2\sigma_1^2)} \Big/ \left(\sigma_1\sqrt{2\pi}\right) \tag{6.2}$$

$$g = e^{-(t-\mu_2)^2/(2\sigma_2^2)} \Big/ \left(\sigma_2\sqrt{2\pi}\right) \tag{6.3}$$

is also another Gaussian function

$$f * g = \frac{1}{\sqrt{2\pi\left(\sigma_1^2+\sigma_2^2\right)}} e^{-[t-(\mu_1+\mu_2)]^2/\left[2\left(\sigma_1^2+\sigma_2^2\right)\right]}. \tag{6.4}$$

where *f*, *g*, and *h* are arbitrary functions and *a* a constant. convolution satisfies the properties:

$$\begin{aligned} & \mathrm{f} * \mathrm{g} = \mathrm{g} * \mathrm{f} \\ & f * (g * h) = (f * g) * h \\ & f * (g + h) = (f * g) + (f * h) \end{aligned}$$

as well as

$$\begin{aligned} a(f * g) &= (af) * g \\ &= f * (ag) \end{aligned}$$

Another property of the convolution integral is:

$$\int_0^t f(t-\tau)\mathrm{g}(\tau)\mathrm{d}\tau = \int_0^t f(\tau)\mathrm{g}(\mathrm{t}-\tau)\mathrm{d}\tau \tag{6.5}$$

The concept of the convolution can be further illustrated by the following example: Consider a dynamic system, in which an input signal, x(t), enters a "black box", S, and the output is y(t) (Fig. 6.1):

Mathematically, we represent the "black box" system as S:

$$\mathrm{y(t)} = \mathrm{S[x(0\,to\,t)]} \tag{6.6}$$

where y(t) represents the output signal at time t, and x(0 to t) represents the time history of the input signal. If the "black box" represents a convolution, it means that

x(t) → S → y(t)

Fig. 6.1 Block diagram illustrating the concept of convolution

S takes on the following mathematical form:

$$y(t) = S[x(0\,\text{to}\,t)] = \int_0^t x(\tau)h_\tau(t-\tau)\delta\tau$$

In this equation $h_\tau(t-\tau)$ is the transfer function (also referred to as the impulse response function). $t-\tau$ is referred to as the lag, as it represents the time lag between the input and the output time. The transfer function is given a subscript τ to indicate that the function itself may take on different forms at different points in time. That is, the response to an input impulse at $t=\tau_1$ could have a different form than the response to an impulse at $t=\tau_2$.

Convolution requires that with the input signal, x(t), and the transfer function, h(t − τ), the convolution integral then computes the output signal, y(t). Note that the convolution integral is a linear operation. That is, for any two functions $x_1(t)$ and $x_2(t)$, and any constant a, the following holds:

$$S[x_1(t) + x_2(t)] = S[x_1(t)] + S[x_2(t)] \tag{6.7}$$

$$S[ax_1(t)] = a\,S[x_1(t)] \tag{6.8}$$

Having defined mathematically what a convolution integral does, the following shows what it represents conceptually. The simplest way is to think about the convolution integral as simply a linear superposition of response functions, $h(t-\tau_i)$, and each of which is multiplied by the impulse $x(\tau_i)\delta\tau$.

A more common way to interpret the convolution integral is to consider the output as a weighted sum of the present and past input values. This can be shown by writing the integral in terms of a sum (and assume here that the system is discretized by a single unit of time):

$$y(t) = x(0)h(t) + x(1)h(t-1) + x(2)h(t-2) + \ldots \tag{6.9}$$

The convolution operation can also be considered as a filtering operation on the signal x(t), where the transfer function is acting as the filter. The shape of the transfer function determines which properties of the original signal x(t) are "filtered out".

6.3 Applications of Convolution

Convolution and related operations have many applications in mathematics, engineering, and science. In digital image processing, convolutional filtering plays an important role in many important algorithms in edge detection and related processes, and in deblurring. In acoustics, reverberation is the convolution of the original sound with echoes from objects surrounding the sound source. In digital signal processing, convolution can be used to map the impulse response of a real room on a digital audio signal. In electronic music, convolution is the imposition of a spectral or rhythmic structure on a sound. Often this envelope or structure is taken from another sound. The convolution of two signals is the filtering of one through the other. In electrical engineering, the convolution of one function (the input signal) with a second function (the impulse response) gives the output of a linear time-invariant system (LTI). At any given moment, the output is an accumulated effect of all the prior values of the input function, with the most recent values typically having the most influence (expressed as a multiplicative factor). The impulse response function provides that factor as a function of the elapsed time since each input value occurred. In physics, wherever there is a linear system with a "superposition principle", the convolution operation will appear. For instance, in spectroscopy line broadening due to the Doppler effect on its own gives a Gaussian spectral line shape and collision broadening alone gives a Lorentzian line shape. When both effects are operative, the line shape is a convolution of Gaussian and Lorentzian, a Voigt function. In computational fluid dynamics, the large eddy simulation (LES) turbulence model uses the convolution operation to lower the range of length scales necessary in computation thereby reducing computational cost. In probability theory, the probability distribution of the sum of two independent random variables is the convolution of their individual distributions. Convolutional neural networks apply multiple cascaded convolution kernels with applications in machine vision and artificial intelligence

6.4 Correlation Function

A correlation function provides the statistical correlation between random variables, contingent on the spatial or temporal distance between those variables. If one considers the correlation function between random variables representing the same quantity measured at two different points then this will become the autocorrelation function. Correlation functions of different random variables are known as cross-correlation functions to emphasize that different variables are being considered.

Correlation functions are a useful indicator of dependencies as a function of distance in space or time, and they can be used to assess the distance required between sample points for the values to be effectively uncorrelated. In addition, they can form the basis of rules for interpolating values at points for which there are no observations.

Correlation functions are commonly used in signal processing and in image processing.

For possibly distinct random variables $X(s)$ and $Y(t)$ at different points s and t of some space, the correlation function is

$$C(s, t) = \mathrm{corr}(X(s), Y(t)) \tag{6.10}$$

This is with the assumption that the random variables are scalar-valued. For the case of vectors, the correlation function will be

$$c_{ij}(\mathrm{s},\mathrm{t}) = \mathrm{corr}\big(X_i(\mathrm{s}), Y_j(\mathrm{t})\big) \tag{6.11}$$

where $X(s)$ is a random vector with n elements and $Y(\mathrm{t})$ is a vector with q elements, and the correlation function is a n × q matrix with l, j elements.

Higher order correlation functions are defined as

$$C_{i_1 i_2 \ldots i_n}(s_1, s_2, \ldots s_n) = \langle X_{i_1}(s_1), X_{i_2}(s_2) \ldots X_{i_n}(s_n) \rangle \tag{6.12}$$

representing a correlation function of order n. If there are symmetries, the complexity can be reduced by breaking up the correlation function into irreducible representations of the internal symmetries and spacetime symmetries.

In fact, the study of correlation functions is similar to the study of probability distributions. For instance, many stochastic processes can be completely characterized by their correlation functions such as the class of Gaussian processes. Probability distributions defined on a finite number of points can always be normalized. However, when defining over continuous spaces, extra conditions are required.

6.5 Autocorrelation

Autocorrelation is defined as the correlation of a signal with a delayed copy of itself as a function of delay. Informally, it is the similarity between observations as a function of the time lag between them. The analysis of autocorrelation is a mathematical tool for finding repeating patterns, such as the presence of a periodic signal obscured by noise, or identifying the missing fundamental frequency in a signal implied by its harmonic frequencies. It is often used in signal processing for analyzing functions or series of values, such as time domain signals.

Let $\{X_t\}$ be a random process, and t be any point in time. Then X_t is the value produced by a given run of the process at time t. Then the definition of autocorrelation between times t_1 and t_2 is

$$R_{XX}(t_1, t_2) = \mathrm{E}\big[X_{t_1} \overline{X_{t_2}}\big] \tag{6.13}$$

where E = expected value operator, and the bar represents complex conjugation.

If $\{X_t\}$ is a broad-sense stationary process and the autocorrelation can be expressed as a function of the time lag $\tau = t_2 - t_1$, then the autocorrelation function can be expressed as

$$R_{XX}(\tau) = \mathrm{E}\big[X_t \overline{X_{t+\tau}}\big] \tag{6.14}$$

R_{XX} being an even function, it has symmetry properties. That is

$$R_{XX}(t_1, t_2) = \overline{R_{XX}(t_2, t_1)} \tag{6.15}$$

and

$$R_{XX}(\tau) = \left[\overline{R_{XX}(-\tau)}\right] \tag{6.16}$$

6.5.1 Autocorrelation of Continuous Time Signal

With a continuous time signal f(t), the autocorrelation $R_{ff}(\tau)$ at lag τ is defined as

$$R_{ff}(\tau) = \int_{-\infty}^{\infty} f(t+\tau)\overline{f(t)}\mathrm{d}t = \int_{-\infty}^{\infty} f(t)\overline{f(t-\tau)}\mathrm{d}t \tag{6.17}$$

where $\overline{f(t)}$ reporesents the complex conjugate of f(t).

6.5.2 Symmetry Property of the Autocorrelation Function

A fundamental property of the autocorrelation function is symmetry. That is

$$R_{ff}(\tau) = R_{ff}(-\tau) \tag{6.18}$$

In the continuous case, the autocorrelation is an even function,

$$R_{ff}(-\tau) = R_{ff}(\tau) \tag{6.19}$$

when f is a real function.

6.5.3 Applications of Autocorrelation

Some examples of the applications of autocorrelation are such as in signal processing, autocorrelation can give information about repeating events like musical beats (for example, to determine tempo) or pulsar frequencies, though it cannot tell the position in time of the beat. It can also be used to estimate the pitch of a musical tone In medical ultrasound imaging, autocorrelation is used to visualize blood flow.

6.6 Cross Correlation

In signal processing, cross-correlation is a measure of the similarity of two time series as a function of the displacement of one relative to the other. It can be used for the searching of a shorter, known feature out of a long signal. It has applications in pattern recognition, electron tomography, and neurophysiology. The cross-correlation is similar in nature to the convolution of two functions. The autocorrelation an be considered as the cross-correlation of a signal with itself. On this case, there will always be a peak at a lag of zero, and its size will be given by the signal energy.

The cross correlation of deterministic signals such as continuous function, it is defined as

$$(\mathrm{f} * \mathrm{g})(\tau) = \int_{-\infty}^{\infty} \overline{f(t)}\mathrm{g}(\mathrm{t} + \tau)\mathrm{dt} \quad (6.20)$$

where $\overline{f(t)}$ denotes the complex conjugate of f(t), and τ is the time lag.

For discrete function, the cross correlation is defined as

$$(\mathrm{f} * \mathrm{g})[\mathrm{n}] = \sum_{-\infty}^{\infty} \overline{f(m)}\mathrm{g}[\mathrm{m} + \mathrm{n}] \quad (6.21)$$

If f and g are two real valued functions differing only by an unknown shift along the x axis, then one can use the cross-correlation to find how much g must be shifted along the x axis to make it identical to f. The formula enables the calculation of the integral of the product of these two functions, with the sliding of the g function along the x axis. The values of (f * g) is maximised when the functions match. This is because for positive areas, when peaks are aligned, they make a large contribution to the integral. When negative areas or troughs are aligned, they also make a positive contribution to the integral, due to the product of two negative numbers is positive. When dealing with complex valued functions f and g, with the conjugate of f, one must ensure that the aligned peaks or aligned troughs with imaginary components will contribute positively to the integral.

6.6.1 *Some Properties of Cross Correlation*

Cross correlation satisfies

$$\mathrm{F}\{\mathrm{f} * \mathrm{g}\} = \overline{F\{f\}} \cdot \mathrm{F}\{\mathrm{g}\} \quad (6.22)$$

where F denotes Fourier transform and $\overline{F\{f\}}$ the complex conjugate of F{f}.

If f and g are both continuous, periodic functions of period T, the limits of integration of integral will be replaced by $[t_0, t_0 + T]$ of length T:

$$(f * g)(\tau) = \int_{t_0}^{t_0+T} \overline{f(t)} g(t + \tau) dt \tag{6.23}$$

For broad sense stationary stochastic process, the cross correlation function has the following symmetry properties:

$$R_{XY}(t_1, t_2) = \overline{R_{YX}}(t_2, t_1) \tag{6.24}$$

and

$$R_{XY}(\tau) = \overline{R_{YX}}(-\tau) \tag{6.25}$$

6.7 Power Spectral Density

From Fourier analysis, one can decompose any physical signal into a spectrum of frequencies or into a number of discrete frequencies over a continuous range. Spectrum is the statistical average of the signal analyzed in terms of its frequency content. The distribution of power into frequency components composing that signal can be described by the power spectrum S_{xx}(f) of a time series x(t). The power spectral density (PSD) is referred to the spectral energy distribution that would be found per unit time. The PSD applies to signals existing over a time period large enough that it could be considered as an infinite time interval. The total energy of such a signal over all time would generally be infinite. Summation or integration of the spectral components yields the total power of a physical process. The power spectrum is important in statistical signal processing. The PSD of the signal describes the power present in the signal as a function of frequency per unit frequency. PSD is usually expressed in watts per hertz. One sometimes uses the amplitude spectral density which is the square root of the PSDF. This is useful only when the shape of the spectrum is rather constant.

The average power P of a signal x(t) over all the time, is a time average given by:

$$P = \lim_{T \to \infty} \frac{1}{T} \int_0^T /x(t)/^2 dt \tag{6.26}$$

A stationary process has a finite power but with an infinite energy. This is because the energy is the integral of power and that the stationary signal can continue over an infinite time. In order to analyze the frequency content of the signal x(t), one has to compute the Fourier Transform X(ω). Due to the fact that the Fourier Transform does not exist for many signals of interest, one has to limit the integral of the signal over a finite interval [0, T] or

$$\mathrm{X}(\boldsymbol{\omega}) = \frac{1}{\sqrt{T}} \int_{0}^{T} x(t) e^{-i\omega t} \mathrm{dt} \tag{6.27}$$

This is the amplitude spectral density. The power spectral density can then be defined as:

$$S_{xx}(\boldsymbol{\omega}) = \lim_{T \to \infty} E\left[/X(\omega)/^{2}\right] \tag{6.28}$$

where E is the expected value.

In terms of the expected value, the autocorrelation function can be written as

$$R_{xx}(\tau) = \mathrm{E}[\mathrm{X(t)}, \mathrm{X}(\mathrm{t} + \tau)] = \langle \mathrm{X(t)X(t} + \tau) \rangle \tag{6.29}$$

The power spectral density in turn is the Fourier Transform of the autocorrelation function:

$$S_{xx}(\boldsymbol{\omega}) = \int_{-\infty}^{\infty} R_{xx}(\tau) \exp(-\mathrm{i}\boldsymbol{\omega}\tau) \mathrm{d}\tau \tag{6.30}$$

This is known as the Wiener-Khinchin Theorem.

6.7.1 Applications of Power Spectral Density

Power spectral density is commonly used in electrical engineering where electronic instruments such as spectrum analyzers are used to measure the power spectra of signals. The concept and use of the power spectrum of a signal is fundamental in electrical engineering, especially in electronic communication engineering such as remote sensing technology, radars and related systems. The spectrum analyser measures the magnitude of the short-time Fourier Transform (STFT) of an input signal. For a stationary process, the STFT will enable a good estimate of the power spectral density.

Reference

1. Wikipedia, Spectral density.

Chapter 7
Wiener Filter and Kalman Filter

7.1 Introduction

In signal processing, Wiener filter is used for noise filtering assuming known stationary signal and noise spectra and additive noise. The Wiener was the first filter for statistical signal processing. It was proposed and designed by Wiener [1]. However, the discrete time form of the filter was independently invented by Kolgomorov [2]. Hence the theory of Wiener filter is also known as the Wiener-Kolmogorov filtering theory. It is used to filter out the additive noise from the corrupted signal to produce an estimate of the underlying signal of interest. The linear time-invariant (LTI) method is used which minimizes the mean square error between the estimated random process and the desired process. This will produce a statistical estimate of an unknown signal as an output with the known signal as the input. The Wiener filter is a form of statistical signal processing.

The design of the Wiener filter takes a different approach from typical deterministic filters designed for a desired frequency response. Here it is assumed that knowledge of the original signal and noise are known. Then one determines the linear time invariant filter whose output would come as closed to the original signal as possible. So the requirements and characteristics of a Wiener filter are:

1. The input signal and additive noise are assumed to be stationary and linear stochastic processes with known autocorrelation and cross-correlation spectral characteristics.
2. The filter must be physically realizable with causal or non-causal solutions.
3. The minimum mean squared error method is used.

W. S. Gan, *Signal Processing and Image Processing for Acoustical Imaging*,
https://doi.org/10.1007/978-981-10-5550-8_7

7.2 Principle of the Wiener Filter

The principle, function, and derivation of the Wiener filter can be illustrated by the following model:

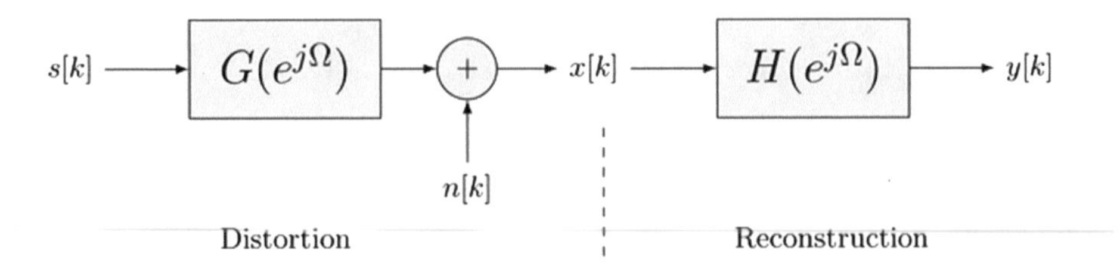

The above model describes that the input random signal s[k] is being distorted by the linear time-invariant system (LTI) system $\mathrm{G}e^{j\Omega}$ and the additive noise n[k] resulting in the output signal as being x[k] = s[k] * g[k] + n[k]. The additive noise n[k] is assumed to be uncorrelated from s[k] and that all random signals are weakly stationary. The distortion model has many practical applications, for instance for the measurement of a physical parameter with a sensor.

The function of the Wiener filter is for designing the LTI system $\mathrm{H}e^{j\Omega}$ with output y[k] matching s[k] as closed as possible.

The error signal e[k] is introduced to quantify this:

$$\mathrm{e[k]} == \mathrm{y[k]} - \mathrm{s[k]} \tag{7.1}$$

The quadratic mean of the error e[k] or the mean squared error (MSE) is then

$$\mathrm{E}\{(e[k])^2\} = \frac{1}{2\pi} \int_{-\pi}^{\pi} \phi_{ee} e^{j\Omega} \mathrm{d}\Omega \tag{7.2}$$

Here one assumes that the cross-power spectral density $\phi_{xs}e^{j\Omega}$ between the observed signal x[k] and input signal s[k] and the power spectral density $\phi_{xx}e^{j\Omega}$ are known. This information is obtained from the estimation from measurements taken at a real system.

The transfer function $\mathrm{H}e^{j\Omega}$ is defined as:

$$\mathrm{H}e^{j\Omega} = \phi_{sx}e^{j\Omega}/\phi_{xx}e^{j\Omega} = \phi_{xs}e^{-j\Omega}/\phi_{xx}e^{j\Omega} \tag{7.3}$$

The optimal Wiener filter is obtained by minimizing the MSE with respect to the transfer function. To obtain the Wiener filter, knowledge of the distortion process is not required.

Fig. 7.1 Image filtering using Wiener filter

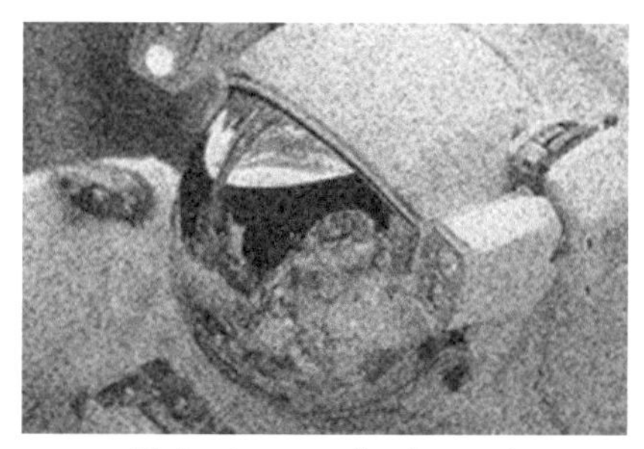

Noisy image of astronaut

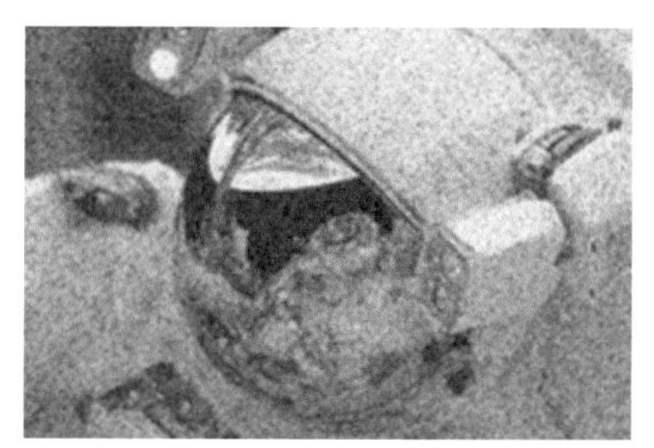

Noisy image of astronaut after Wiener filter applied

7.3 Applications

To implement the Wiener filter in practice, for instance in image processing, one has to estimate the power spectra density of the original image to be filtered and the additive noise. For white additive noise, the power spectrum density is equal to the variance of the noise.

The Wiener filter has various applications in signal processing and image processing in the areas of deconvolution, noise reduction, signal reduction, and system identification.

In the application in signal processing, it is can be used to denoise audio signals, especially speech, as a preprocessor before speech recognition. In the application in image processing to remove noise from a picture producing the filtered image below (see Fig. 1):

Noisy image of astronaut.

Noisy image of astronaut after Wiener filter applied.

7.4 Kalman Filter-Introduction

Kalman filtering is an algorithm used in signal processing that uses a series of measurements observed over time, containing statistical noise and other inaccuracies, instead of a single measurement. It produces estimates of unknown variables that tend to be more accurate than those based on a single measurement. This is done by estimating a joint probability distribution over the variables for each time frame.

Kalman filter is names after Kalman [3], one of the primary developers of the theory. It is very useful in signal processing and has several technological applications related to aircrafts, spacecrafts, ships, robots etc. The algorithm of the Kalman filter works in two steps. First estimates of the current state variables with their uncertainties are produced The next measurement is observed corrupted with some error, including random noise. These estimates are then updated using a weighted average. Here more weight is given to estimates with higher certainty. So this is a recursive algorithm. It can be run in real time. Only the present input measurements, previously calculated state and its uncertainty matrix are needed.

Kalman filter does not have to assume that the errors are Gaussian. However for the special case that all errors are Gaussian, the filter will produce the exact conditional probability estimate.

7.5 The Algorithm of the Kalman Filter

Kalman filter is a very useful mathematical tool for stochastic estimate from noisy sensor measurements. It is a set of mathematical equations that implement a predictor-corrector type estimator that is optimal in the sense that it minimizes the estimated error covariance when presumed conditions are met. With the advance in digital computing, the Kalman filter has become more practical in use with relative simplicity and robust nature.

The original Kalman filter formulated by Kalman in 1960 is a discrete filter. The measurements and the state are estimated at discrete points in time.

One will try to estimate the state x $\in R^n$ of a discrete time controlled process that is governed by a linear stochastic difference equation

$$x_k = \mathrm{A}x_{k-1} + \mathrm{B}u_k + w_{k-1} \tag{7.4}$$

with a measurement $z_k = \mathrm{H}x_k + v_k$, $w_k =$ process noise, and $v_k =$ measurement noise.

Kalman filter estimates a process by using a form of feedback control. The filter estimates the process state at some time and then obtain feedback in the form of noisy measurements. As such, equations of Kalman filter fall into two groups: time update equations and measurement update equations. Time update equations projecting forward in time the current state and error covariance estimates to obtain a priori estimates for next time step. The measurement update equations are responsible for the feedback, for incorporating a new measurement into the a priori estimate to obtain an improved a posteriori estimate. Here one considers the time update equations as predictor equations, and the measurement update equations as corrector equations.

Fig. 7.2 The Discrete Kalman filter cycle. The time update projects the current state estimate ahead in time. The measurement update adjusts the projected estimate by an actual measurement at that time

The final estimation algorithm will resemble that of a predictor-corrector algorithm for solving numerical problems. This is shown in Fig. 7.2.

The specific equations for the time and measurement updates are shown below.

Discrete Kalman filter time update equations:

$$\widehat{x_k}^{-} = \mathrm{A}\widehat{x_{k-1}} + \mathrm{B}u_k \tag{7.5}$$

$$P_k^{-} = \mathrm{A}P_{k-1}A^T + \mathrm{Q} \tag{7.6}$$

Discrete Kalman filter measurement update equations:

$$K_k = P_k^{-}H^T\left(\mathrm{H}P_k^{-}H^T + R\right)^{-1} \tag{7.7}$$

$$\widehat{x_k} = \widehat{x_k}^{-} + k_k\left(Z_k - \mathrm{H}\widehat{x_k}^{-}\right) \tag{7.8}$$

$$P_k = (\mathrm{I} - K_k\mathrm{H})P_k^{-} \tag{7.9}$$

The time update equations project the state and covariance estimates forward from time step k − 1 to step k. A and B are from Eq. (7.4).

The first task during the measurement update is to compute the Kalman gain K_k. The next step is to measure the process z_k to obtain, and then to generate an a posteriori state estimate by incorporating the measurement as in Eq. (7.8). The final step is to obtain an a posteriori error covariance estimate via Eq. (7.9).

After each time and measurement update pair, the process is repeated with the previous a posteriori estimates used to project or predict the new a priori estimates. This recursive nature is one of the very appealing features of the Kalman filter—it makes practical implementations much more feasible than (for example) an implementation of a Wiener filter which is designed to operate on all of the data directly for each estimate. The Kalman filter instead recursively conditions the current estimate on all of the past measurements.

References

1. Wiener, N. (1949). *Extrapolation, interpolation, and smoothing of stationary time series*. New York: Wiley.
2. Kolmogorov, A. N. (1941). Stationary sequences in Hilbert space. *Bulletin Moscow University Mathematics*, 2(6), 1–40. (in Russian). English Trans. in Kailath, T. (Ed.). (1977). *Linear least squares estimation*. Dowden, Hutchinson and Ross.
3. Kalman, R. E. (1960). A new approach to linear filtering and prediction problems. *Transaction of the ASME—Journal of Basic Engineering, 82*(Series D), 34–35.

Chapter 8
Higher Order Statistics

8.1 Introduction

Higher order statistics (HOS) are meant for nonlinear signal processing, dealing with signals from nonlinear systems. They are used to extract information from random signals. Second order statistics like autocorrelation function and probability density function are used only for linear signals such as Gaussian noise. Nonlinear signal processing can be applied to pattern recognition and communication. Some examples of nonlinear signals are music, speech signals, EEG, and ECG signals. HOS will enable better performance from sensors, sensor networks and channels with applications in filtering, detection and coding. HOS can be applied to both deterministic signals and to stochastic framework. The earliest work on HOS is that of Brillinger and Rosenblatt [1, 2] which laid the theoretical foundation. Subsequently there was the paper of Hasselman et al. [3] that applied bispectra to study ocean waves. It was not until the 1970s that HOS is applied to real signal processing problems. In recent years, applications have been diversified to as broad as to speech, seismic data processing, acoustical image processing, optics, plasma physics, and economics.

In the time domain, the second order statistics is the autocorrelation function $\mathrm{R(m)} = \langle \mathrm{x(n)x(n + m)} \rangle$ where $\langle\rangle$ = expectation operator. The third order statistics is the third-order moment The third order statistics is the third order moment: $\mathrm{M(m1, m2)} = \langle \mathrm{x(n)(n + m1)x(n + m2)} \rangle$. Note that the third order moment depends on two independent lags m1 and m2. This can be extended into the fourth-order moment and higher order moments by adding on lag terms to the above equations.

The Fourier transform of the third order cumulant function is referred to as the bispectrum and it is a complex valued function of two frequencies:

$$\mathrm{B}(f_1, f_2) = \int_{-\infty}^{\infty} R(\tau_1, \tau_2) \exp\left[-\mathrm{j}2\pi(f_1\tau_1 + f_2\tau_2)\mathrm{d}\tau_1\mathrm{d}\tau_2\right]. \tag{8.1}$$

W. S. Gan, *Signal Processing and Image Processing for Acoustical Imaging*,
https://doi.org/10.1007/978-981-10-5550-8_8

The Fourier transform of the fourth order cumulant function is referred to as the trispectrum. It may be noted that these Fourier spectra exist only when the cumulant functions are well behaved. That is they decay with increasing lag and are absolutely integrable.

The third order correlation or third order moment is a function of two lag variables and the bispectrum is a function of two frequency variables. It is possible to estimate the bispectrum in two ways: either computing the third moment function and then taking its Fourier transform or computing the Fourier transform of the function and using triple products of Fourier coefficients.

The Fourier transform of the 4th order cumulant function is referred to as the trispectrum.

If {x(k)} with k = 0, ±1, ±2, ±3, ... is a real stationary discrete time signal and its moments up to order n exist, then

$$m_n^x(\tau_1, \tau_2, \tau_3, \ldots \tau_{n-1}) \underline{\Delta} \mathrm{E}\big[\mathrm{x(k)}, \mathrm{x}\big(\mathrm{k} + \tau_1 \mathrm{x}(\mathrm{k} + \tau_2) \ldots \mathrm{x}(\mathrm{k} + \tau_{n-1} 0\big] \quad (8.2)$$

represents the nth order moment function of the signal, which depends only on the time difference $\tau_1, \tau_2, \ldots \tau_{n-1,} \tau_i = 0, \pm 1, \pm 2, \ldots$ for all i. The second, moment function $m_2^x(\tau_1)$ is the autocorrelation whereas $m_3^x(\tau_1, \tau_2)$ and $m_4^x(\tau_1, \tau_2, \tau_3)$ are the 3rd, 4th order moments respectively. They are the 3rd and 4th autocorrelation of the signal. The nth order cumulant of a non-Gaussian stationary random signal x(k) can be written as (for n = 3, 4 only):

$$c_n^x(\tau_1, \tau_2, \ldots \tau_{n-1}) = m_n^x(\tau_1, \tau_2, \ldots \tau_{n-1}) - m_n^G(\tau_1, \tau_2, \ldots \tau_{n-1}) \quad (8.3)$$

where $m_n^x(\tau_1, \tau_2, \ldots \tau_{n-1})$ is the nth moment function of x(k) and $m_n^G(\tau_1, \tau_2, \ldots \tau_{n-1})$ is the nth order moment function of an equivalent Gaussian signal that has the same mean value and autocorrelation sequence as x(k). Clearly, if x(k) is Gaussian, $m_n^x(\tau_1, \tau_2, \ldots \tau_{n-1}) = m_n^G(\tau_1, \tau_2, \ldots \tau_{n-1})$ and thus $c_n^x(\tau_1, \tau_2, \ldots \tau_{n-1}) = 0$. Note, however, that although Eq. (8.3) is only true for orders n = 3 and 4, $c_n^x(\tau_1, \tau_2, \ldots \tau_{n-1}) = 0$ for all n if x(k) is Gaussian.

If the signal x(k) is zero means $m_1^x = 0$, it follows from (8.4) and (8.5) that the second and the third order cumulants are identical to the second and third moments respectively. The third order autocorrelation function is the third order moment.

Equation (8.3) is only true for orders n = 3 and 4.

$$c_2^x(\tau_1) = m_2^x(\tau_1) - \left(m_1^x\right)^2 = m_2^2(-\tau_1) - \left(m_1^x\right)^2 = m_2^2(-\tau_1) \quad (8.4)$$

where $m_2^2(-\tau_1)$ is the autocorrelation sequence. Thus, we see that the second order cumulant sequence is the covariance while the second order moment sequence is the autocorrelation.

The third order cumulant is given by

$$c_3^x(\tau_1, \tau_2) = m_x^3(\tau_1, \tau_2) - m_1^x\left[m_2^x(\tau_1) + m_2^x(\tau_2) + m_2^x(\tau_1 - \tau_2)\right] + 2\left(m_1^x\right)^3 \quad (8.5)$$

where $m_x^3(\tau_1, \tau_2)$ is the third order moment sequence.. This follows if we combine (8.3) and (8.4).

The fourth order cumulant: if we combine (8.3) and (8.4), we will obtain

$$\begin{aligned} c_4^x(\tau_1, \tau_2, \tau_3) &= m_4^x(\tau_1, \tau_2, \tau_3) - m_2^x(\tau_1)m_2^x(\tau_3 - \tau_2) - m_2^x(\tau_2)m_2^x(\tau_3 - \tau_1) \\ &- m_2^x(\tau_3)m_2^x(\tau_2 - \tau_1) - m_1^x\left[m_3^x(\tau_2 - \tau_1, \tau_3 - \tau_1) + m_x^3(\tau_2, \tau_3) + m_3^x(\tau_3, \tau_1) + m_3^x(\tau_1, \tau_2)\right] \\ &+ \left(m_2^x\right)^2\left[m_1^x(\tau_1) + m_2^x(\tau_2) + m_2^x(\tau_3) + m_2^x(\tau_3 - \tau_1) + m_2^x(\tau_3 - \tau_2) + m_2^2(\tau_2 - \tau_1)\right] - 6\left(m_1^x\right)^4 \end{aligned} \tag{8.6}$$

If the signal {x(k)} is zero means $m_1^x = 0$, it follows from (8.4) and (8.5) that the second and third order cumulants are identical to the second and the third order moments respectively. However, to generate the fourth order cumulants, we need knowledge of the fourth order and second order moments in Eq. (8.6), that is

$$c_4^x(\tau_1, \tau_2, \tau_3) = m_4^x(\tau_1, \tau_2, \tau_3) - m_2^x(\tau_1)m_2^x(\tau_3 - \tau_2) - m_2^x(\tau_2)m_2^x(\tau_3 - \tau_1) - m_2^x(\tau_3)m_2^x(\tau_2 - \tau_1) \tag{8.7}$$

Reference

1. White, P. (2009). Encyclopedia of structural health monitoring. In *Higher order statistical signal processing*. Wiley.
2. Brillinger, D. R., & Rosenblatt, M. Asymptotic theory of estimates of k-th order spectra. In *Spectral analysis of time series*, B. Harris Ed. New York: Wiley, 1967, pp. 153–188.
3. Hasselman, K., Munk, W., & MacDonald, G. Bispectra of ocean waves, in Time Series Analysis, M. Rosenblatt, Ed. New York: Wiley, 1963, pp. 125–130.
4. Brillinger, D. R., & Rosenblatt, M. Computation and Interpretation of k-th order spectra. In *Spectral analysis of time series*, B. Harris, Ed. New York: Wiley, 1967b, pp. 189-232.

Chapter 9
Digital Signal Processing

9.1 Introduction

Digital signal processing [1] is an important branch of signal processing. It has wide applications such as in audio and speech processing, sonar, radar, sensor array processing, statistical signal processing, seismology, control systems, telecommunication, biomedical imaging, and so on. Digital signal processing will include both linear and nonlinear systems, operating in time, frequency, and spatio-temporal domains. The advantages of digital signal processing over the analog processing are in the areas of error detection and correction in transmission and in data compression.

9.2 Signal Sampling

To perform digital signal processing, the first step the signal has to be digitized with an analog-to-digital converter (ADC). This is known as sampling. This is performed in two stages: discretization and quantization. Discretization means the division of signal into equal intervals of time with each interval represented by a single measurement of amplitude. Quantization is the approximation of each amplitude measurement by a value from a finite set.

The well known Nyquist-Shannon sampling theorem states that a signal can be exactly reconstructed from its samples if the sampling frequency is greater than twice the highest frequency component in the signal. The Nyquist frequency is half of the sampling rate of a discrete signal processing system.

Theoretical digital signal processing works are based on discrete time signal modes which are analysis and derivations with no amplitude inaccuracies created by the abstract process of sampling or quantization error. The signals produced by the ADC and the digital-to-analog converter (DAC) are quantized signals.

W. S. Gan, *Signal Processing and Image Processing for Acoustical Imaging*,
https://doi.org/10.1007/978-981-10-5550-8_9

9.3 Domains of Digital Signal Processing

There are various domains in the study of digital signal processing such as the time domain, frequency domain, and spatial domain. The required domain to be chosen depends on the characteristics of the signal and the processing method to be used. For instance, a discrete Fourier transform will require a frequency representation and the sequence of samples from a measuring device will fit into a spatial domain representation.

9.3.1 Time Domain

The common method used in digital signal processing in the time domain is filtering. This is an enhancement of the input signal. Digital filtering is a form of linear transformation of a number of samples surrounding around the current sample of the input or output signal. The following shows the various types of filters:

1. A linear filter: this is a linear transformation of input samples which follows the superposition principle which states that if the output is a weighted linear combination of various signals, then the output will be a similarly weighted linear combination of the corresponding output signals.
2. A finite impulse response (FIR) filter which uses only the input signals. FIR filters are always stable.
3. An infinite impulse response (IIR) filter which uses both the input signal and previous examples of the output signal. IIR filters may be unstable.
4. A time-invariant filter which has constant property over time.
5. An adaptive filter which has properties change in time.
6. A stable filter which produces an output that converges to a constant value with time or remain bounded within a finite interval.
7. An unstable filter which produces an output that grows without bounds, with bounded or even zero input.
8. A causal filter which used only the previous samples of the input or output signals.
9. A non-causal filter uses future input samples. It can be changed into a causal filter by adding on a delay to it.

The function of a filter can be represented by a block diagram which can be used to derive the signal processing algorithm to implement the filter with hardware components. A difference equation with a collection of zeros and poles or a step response or impulse response can also be used to describe a filter. Hence the output of the linear digital filter to any given input can be calculated by convolving the input signal with the impulse response. By convolving the input signal with the impulse response, one can obtain the output of a linear digital filter.

9.3.2 *Frequency Domain*

Fourier transform is used to convert signals from the time or space domain to frequency domain. The time or space information of the signal is being converted into the magnitude and phase component of each frequency. This can provide information on how the phase varies with frequency. Also the magnitude of each frequency component squared can provide the power spectrum.

The purpose of signal processing in the frequency domain is the analysis of signal properties. The frequency spectrum can be studied to determine which frequencies are present in the input signal and which are missing. The analysis of signals in the frequency domain is also known as spectrum analysis.

For non-real time work, filtering can also be done in the frequency domain, and then converting back to the time domain. This is an efficient implementation and can give essentially a filter response.

An example of a commonly used frequency domain transformation is the cepstrum which converts a signal to the frequency domain through Fourier transform. Then taking the logarithm, and applying another Fourier transform. This can emphasize the harmonic structure of the original spectrum.

9.4 Useful Tools in Digital Signal Processing

9.4.1 *Wavelet Transform*

The discrete wavelet transform (DWT) is usually used in digital signal processing. Here the wavelets are discretely sampled. The advantage of wavelet transform over Fourier transform is temporal resolution which captures both frequency and location information. The accuracy of this joint time-frequency resolution is limited by the uncertainty principle of time-frequency.

9.4.2 *Z-Transform*

There are two forms of digital filters: the IIR type and the FIR type. FIR filters are always stable, whereas IIR filters have feedback loops that can become unstable and oscillate. The Z-transform is a tool for analyzing the stability issues of the digital IIR filters. It is analogous to Laplace transform used in the design and the analysis of analog IIR filters.

9.5 The Implementation of Digital Signal Processing

For systems that do not require real-time computing and if the input or output signal data exist in data files, then the digital signal processing algorithm can be run economically on general purpose computers. This is just ordinary data processing, except that the sampled data is assumed to be uniformly sampled in time or space with the use of DSP mathematical techniques such as the FFT.

When real time application is required, then specialised or dedicated microprocessors or processors have to be used. These will process data using floating point or fixed-point arithmetic. For even more demanding requirements, field-programmable gate array (FPGAs) have to be used. For the most demanding situation, application-specific integrated circuit (ASICs) have to be designed specifically for the application.

9.6 Applications of Digital Signal Processing

The following are the frequent applications of digital signal processing: digital image processing, audio signal processing, speech processing, speech recognition, audio compression, seismology, biomedicine, speech coding and transmission, sonar, digital synthesizers, room correction of sound in hi-fi and sound reinforcement, audio crossover and equalization, ultrasound imaging etc.

9.7 Digital Signal Processor

A digital signal processor (DSP) is a specialized microprocessor with its architecture optimized for the operational needs of digital signal processing. There are two types: the general purpose type and the dedicated type. Most of the general purpose types can execute digital signal processing successfully but may not be able to keep up with such processing continuously in real-time. The dedicated DSPs are more power efficient and thus more suitable to be used in portable devices such as mobile phones. DSPs have special memory architectures for multiple data or simultaneous instructions.

The architecture of a DSP is optimized specifically for digital signal processing and also as a microcontroller or as an application processor.

A DSP is a hardware and a chip. Real world signals like audio, speech, pressure, position, temperature, video are analog in nature and have to be digitized and then mathematically manipulated. The DSP is designed to perform mathematical functions like addition, subtraction, multiplication and division at a high speed. Analog signals have to be processed so that the informations they contain can be displayed, analysed or converted into another form of signal that can be of practical use.

Analog products detect analog signals. These real world signals are converted into digital formats of 1's and 0's. The DSP then captures these digitized informations and process them. These digitized informations are then feeded back for use in the real world. This is done at very high speeds either digitally or in an analog format by going through a digital-to-analog converter. This procedure is illustrated by an example of a DSP used in a MP3 audio player. During the recording phase, the analog audio is in out through a receiver. This analog signal is then converted into a digital signal by the analog-to-digital converter and passed on to the DSP which performs the MP3 encoding and saves the file to memory. The next step is the playback phase, with the file taken from memory, then decoded by the DSP and then converted back into an analog signal using the digital-to-analog converter and output through the speaker system.

The applications of a DSP are the information from a DSP can be used by a computer to control parameters such as security, home theatre system, telephone, and video compression. With a DSP, signals can be compressed so that they can be transmitted more efficiently and with high speed from one location to another. For instance, teleconferencing can transmit speed and video via telephone lines. Also signals can be manipulated or enhanced to improve theory quality, for instance, computer-enhanced medical images and echo cancellation for cell phones. Digital signal processing provides the advantages of accuracy and high speed compared with analog signal processing. A DSP is programmable and so it can be used in a wide range of applications. One can create own software to design a DSP solution to an application (Fig. 9.1).

The key components of a DSP are:

a. Program Memory: this stores the programs and will be used to process data.
b. Data Memory: this stores the information to be processed.
c. Compute Engine: this performs the math processing, accessing the program from the Program Memory and the Data Memory.
d. Input/Output: this consists of a range of functions to connect to the outside world.

Fig. 9.1 MP3 audio player

9.7.1 Architecture of a Digital Signal Processor

Software Architecture

The software structure of a digital signal processor consists hand-optimized assembly-code routines. Commonly packaged into libraries for re-use. Hand-optimized assembly code is more efficient even with modern compiler optimizations. Hence many common algorithms involved in DSP calculations are hand-written.

Hardware Architecture

The hardware architecture of a digital signal processing system refers to the identification of the system's physical components and their interrelationships. This is often called a hardware design model which allows hardware designers to understand how their components fit into a system architecture and provides to software component designers important information needed for software development and integration. There is the need to effectively model how separate physical components combined to form complex system.

Reference

1. Rabiner, L. R., & Gold, B. (1975). *Theory and application of digital signal processing*. Englewood Cliffs, NJ: Prentice-Hall Inc.

Chapter 10
Digital Image Processing

10.1 Background

Digital image processing [1] is the mathematical processing of digital images. Computer algorithms are used. It has several advantages over analog image processing such as allowing a much wider range of algorithms to be applied to the input data and the problems such as the build-up of noise and signal distortion during processing can be avoided. As an image can be considered as a two dimension and multi-dimensional signals, digital image processing can cover multidimensional systems.

Digital image processing techniques were discovered in the 1960s with applications to satellite imagery, medical imaging, character recognition, videophone and so on. The processing cost was fairly high due to the computing price of that era. The situation changed in the 1970s when cheaper computers and dedicated hardware become available. The speed of general-purpose computers became higher and started to take over the role of dedicated hardware. In the 2000s, fast computers and signal processors became available and digital image processing became the most common and the cheapest and the most versatile form of image processing. Image is considered as a two dimension and higher dimensions signal. So many of the techniques used in signal processing are also used in digital image processing. Digital image processing includes these three basic steps: inputing the digital image via image acquisition tools, analysing and manipulating the image and the output which can be the improved image or report based on image analysis.

W. S. Gan, *Signal Processing and Image Processing for Acoustical Imaging*,
https://doi.org/10.1007/978-981-10-5550-8_10

10.2 Scope of Work

Complex algorithms used in digital image processing enable more sophisticated performance of simple tasks and the use of methods impossible by analog means. Examples are the classification of images, pattern recognition, projection, feature extraction, and multi-scale signal analysis. Some techniques used in digital image processing are: partial differential equation, hidden Markov models, linear filtering, image restoration, image editing, neural networks, wavelets, and self-organizing maps.

10.3 Some Procedures of Digital Image Transformations

10.3.1 Image Filtering

Digital filters can be used to deblur and sharpen digital images. The filtering procedures are: convolution with specifically designed kernels or filter array in the spatial domain and masking specific frequency regions in the frequency or Fourier domain.

An image can be defined as a two-dimensional function such as F(x,y) where x and y are the spatial coordinates and the square of the amplitude of F at any pair of coordinates (x,y) is known as the image intensity at that point. When x,y and the amplitude values of F are finite, we call it a digital image. This means that an image can be defined by a two-dimensional array specifically arranged in rows and columns. A digital image is composed of a finite number of elements, each of which elements has a particular value at a particular location. These elements are termed as pixels.

10.4 Types of Digital Images

10.4.1 Binary Image

The binary image consists of only two pixel elements 0 and 1 where 0 refers to black and 1 refers to white. The image is also known as monochrome.

10.4.2 Black and White Image

The image consists only of black and white colour and is also known as black and white image.

10.4.3 8 Bit Colour Format

This is the most common image format with 256 different shades of colours and is also commonly known as the gray scale image. In this format, 0 stands for black, stands for white, and 127 stands for gay.

10.4.4 16 Bit Colour Format

It is a colour image format containing 65,536 different colours, and is also known as high colour format. In this format the distribution of colour is not the same as gray scale image. A 16 bit format is actually divided into three further formats known as the red, green, and blue. This is the RGB format.

10.5 Image as a Matrix

Digital images are represented in rows and in columns as shown in the following syntax in which images are represented.

$$\mathrm{f(x,\,y)} = \begin{bmatrix} f(0,0) & f(0,1)\ldots & f(0,N-1) \\ f(1,0) & f(1,1)\ldots & f(1,N-1) \\ f(M-1,0)\ldots & f(M-1,1) & f(M-1,N-1) \end{bmatrix} \tag{10.1}$$

The right hand side of this matrix is digitized image by definition. Every element of this matrix is the image element or pixel. The following is the digital image representation in MATLAB.

$$\mathrm{f(x,\,y)} = \begin{bmatrix} f(1,1) & f(1,2)\ldots & f(1,N) \\ f(2,1) & f(2,2)\ldots & f(2,N) \\ f(M,1) & f(M,2)\ldots & f(M,N) \end{bmatrix} \tag{10.2}$$

In MATLAB, the Index is starting from 1 instead of from 0. Therefore, f(1,1) $\rightarrow$ f(0,0). The two representations of images are identical, except for the shift in origin. In MATLAB, matrices are stored in a variable.

10.6 Scope of Image Processing

10.6.1 Acquisition

Here an image is given in a digital format. The works involve scaling, and colour conversion from RGB to gray or vice versa.

10.6.2 Image Enhancement

This is used for the extraction of some hidden details from an image and to increase the clarity of the image.

10.6.3 Image Restoration

This deals with restoring the missing information on the image. It is based on mathematical model.

10.6.4 Colour Image Processing

This deals with pseudocolour and full colour image processing. Colour models are applicable to digital image processing.

10.6.5 Wavelets and Multi-resolution Processing

This is the foundation of the representation of images in various degrees.

10.6.6 Image Compression

Mathematical functions have to be developed to perform this operation. It deals with the image size and the image resolution.

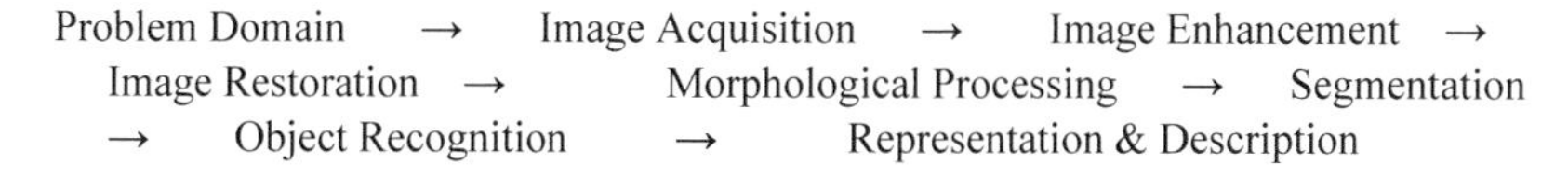

Fig. 10.1 Block diagram of digital image processing

10.6.7 Morphological Processing

This deals with the extraction of image components that are useful in the representation and description of shape. Mathematical tools have to be used.

10.6.8 Segmentation Procedure

This includes the partitioning of an image into its constituent parts or objects.

10.6.9 Representation and Description

This follows the output of segmentation stage.

10.7 Key Stages in Digital Image Processing

This can be shown by the following block diagram (Fig. 10.1).

Reference

1. Gonzalez, R. C., & Woods, R. E. (2002). *Digital image processing*. Boston: Addison-Wesley.

Chapter 11
Digital Image Filtering

11.1 Digital Filtering

Images are distorted by noise. So image filtering basically is the filtering of noise from the images. In digital filtering, the filters play a key role. Filters re-evaluate the value of each pixel in an image. For a particular pixel, the new value is based of pixel values in a local neighbourhood, a window centred on that pixel, in order to reduce noise by smoothing, and/or enhance edges. Filters are an aid to visual interpretation of images, and can also be used as a precursor to further digital processing, such as segmentation. They can be applied directly to record images or after transformation of pixel values. A digital image can be filtered either in the frequency domain or in the spatial domain. For filtering in the frequency domain, first the digital image has to be transformed into the frequency domain. Then it has to be multiplied with the frequency filter function and re-transformed the result back into the spatial domain. The frequency filter function has to be shaped so as to attenuate some frequency and enhance the others. For example, a simple low pass function is 1 for frequencies smaller than the cut-off frequency and 0 for all others.

For filtering in the spatial domain, the procedure is convolution of the input image f(i, j) with the filter function h(i, j):

$$G(i, j) = h(i, j) * f(i, j) \tag{11.1}$$

The filter function will be approximated by a discrete and finite matrix. The convolution will be a shift and multiply operation by shifting the matrix over the image and multiply its value by the corresponding pixel values of the image. For a square matrix of size $M \times M$, one will calculate the output image using the following equation:

W. S. Gan, *Signal Processing and Image Processing for Acoustical Imaging*,
https://doi.org/10.1007/978-981-10-5550-8_11

$$\text{g(i, j)} = \sum_{m=-M/2}^{M/2} \sum_{n=-M/2}^{M/2} h(m, n) f(i - m, j - n) \quad (11.2)$$

Different specific applications will require different forms of matrices.

11.2 Convolution

Convolution is an important procedure in digital image filtering [1] in the sharpening and removing of noise in image processing.

The digital filter is generated by providing a set of weights to apply to the corresponding pixels in a given size neighbourhood. This set of weights will form what is known as the convolution kernel which is represented as a matrix with position of elements correspond to the appropriate pixel in the neighbourhood. This digital filter in the form of a convolution kernel is usually a square matrix with odd dimension so that when applied, the resulting image, will not shift a half-pixel relative to the original image. The following shows the general form of a 3×3 convolution kernel:

$$\begin{bmatrix} w1 & w2 & w3 \\ w4 & w5 & w6 \\ w7 & w8 & w9 \end{bmatrix}$$

If the weights are not designed to sum to unity then the following normalization has to be done:

$$\frac{1}{sum\ w} \begin{bmatrix} w1 & w2 & w3 \\ w4 & w5 & w6 \\ w7 & w8 & w9 \end{bmatrix}$$

where sum w = w1 + w2 + w3 + w4 + w5 + w6 + w7 + w8 + w9.

This normalization factor must be taken into account if the filter is mixed with other filters to generate a more complex convolution.

The simplest filter or convolution kernel is of size 3×3 with equal weights shown as follows:

$$\begin{bmatrix} 1/9 & 1/9 & 1/9 \\ 1/9 & 1/9 & 1/9 \\ 1/9 & 1/9 & 1/9 \end{bmatrix} = \frac{1}{9} \begin{bmatrix} 1 & 1 & 1 \\ 1 & 1 & 1 \\ 1 & 1 & 1 \end{bmatrix} = \text{low pass filter}$$

The above figure is an example of a low pass filter. It produces the arithmetic mean of the 9 nearest neighbours of each pixel in the image. This type of filter passes the low frequencies or long wavelengths in the image. Conversely, high frequencies or short wavelengths are removed. This will give rise to abrupt changes in image intensities, i.e.edges, resulting in blurring. In addition, due to replacement of each pixel by an average of the pixels in its local neighbourhood, it tends to blur the image.

High pass filter is the opposite of a low pass filter. It passes only the high frequencies or short wavelengths. When applied to an image, it shows only the edges in the image.

Usually high pass filtering is achieved by the application of a low pass filter to the image and the subsequent subtraction of the low pass filtered result from the image. This can be represented by $H = I - L$ where $H =$ high pass filtered image, $L =$ low pass filtered image, and $I =$ original image or in terms of the filters (convolution kernels) as

$$h\Theta I = i\Theta I - l\Theta I \tag{11.3}$$

where h, i, and l are the high pass, identity, and low pass filter (convolution kernels) respectively and Θ stands for convolution or multiplying each neighbourhood pixel by its corresponding weight and summing up the products.

The identity kernel is a convolution kernel which when applied to the image leaves the image unchanged. It is illustrated as follows:

$$i = \begin{bmatrix} 0\ 1\ 0 \\ 0\ 1\ 0 \\ 0\ 0\ 0 \end{bmatrix} = \text{identity kernel} \tag{11.4}$$

Combining the identity kernel and low pass filter kernel, one has

$$h = \begin{bmatrix} 0\ 0\ 0 \\ 0\ 1\ 0 \\ 0\ 0\ 0 \end{bmatrix} - \frac{1}{9}\begin{bmatrix} 1\ 1\ 1 \\ 1\ 1\ 1 \\ 1\ 1\ 1 \end{bmatrix} = \frac{1}{9}\begin{bmatrix} -1 & -1 & -1 \\ -1 & 8 & -1 \\ -1 & -1 & -1 \end{bmatrix} = \text{high pass filter}$$

Actually the one-ninth factor can be ignored because the sum of weights is zero which is a typical property of pure high pass filter.

11.3 Application of Digital Filters in Digital Image Processing

An example is the extraction of edges in the image. This can be done by the high pass filters. Next one will show that the high pass filter can also be used for the sharpening of images. This can be shown as

$$S = (1 - f)^{*}I + f^{*}H \tag{11.5}$$

where S = sharpened image, I = original image, H = high pass filtered image, f = a fraction between 0 and 1.

When f = 0, S = I, and when f = = 1, S = H. If one takes f = 0.5 as something in between, then

$$S\Theta I = (1 - f)^* i\Theta I + f^* h\Theta I \tag{11.6}$$

$$\text{giving} \;\; S = (1 - f)^* I + f^* h \tag{11.7}$$

For the high pass filter, this will become

$$S = (1 - f)\begin{bmatrix} 0 & 0 & 0 \\ 0 & 1 & 0 \\ 0 & 0 & 0 \end{bmatrix} + f\frac{1}{9}\begin{bmatrix} -1 & -1 & -1 \\ -1 & 8 & -1 \\ -1 & -1 & -1 \end{bmatrix}$$
$$= \frac{1}{9}\begin{bmatrix} -f & -f & -f \\ -f & (9 - f) & -f \\ -f & -f & -f \end{bmatrix}$$

The following example in Fig. 11.1 will illustrate the 3 × 3 uniform weight (averaging) convolution kernels as filters.

11.4 Digital Linear Filter in the Spatial Domain

Linear methods are for more amenable to mathematical analysis than are nonlinear ones and are consequently far better understood. Linear digital filters can be operated both in the spatial domain and in the frequency domain. Linear filters are filters whose output values are linear combinations of the pixels in the original image.

The simplest linear digital filter is the moving average filter. Here each pixel is replaced by the average of pixel values in a square centred at that pixel. Let f_{ij}, for i, j = 1, …, n, denote the pixel values in the image as input and g_{ij} be the pixel values of the output. With a linear filter of size (2m + 1) × (2n + 1) with weights w_{kl} for k, l = −m, …, m, then

$$g_{ij} = \sum_{k=-m}^{m}\sum_{l=-m}^{m} w_{kl} f_{i+k,j+l} \tag{11.8}$$

for i, j = (m + 1), …, (n − m).

If m = 1, then the window over which averaging is carried out is 3 × 3, giving,

$$g_{ij} = w_{-1,-1}f_{i-1,j-1} + w_{-1,0}f_{i-1,j} + w_{-1,1}f_{i-1,j+1} + w_{0,-1}f_{i,j-1} + w_{0,0}f_{i,j}$$
$$+ w_{0,1}f_{i,j+1} + w_{1,-1}f_{i+1,j-1} + w_{1,0}f_{i+1,j} + w_{1,1}f_{i+1,j+1}$$

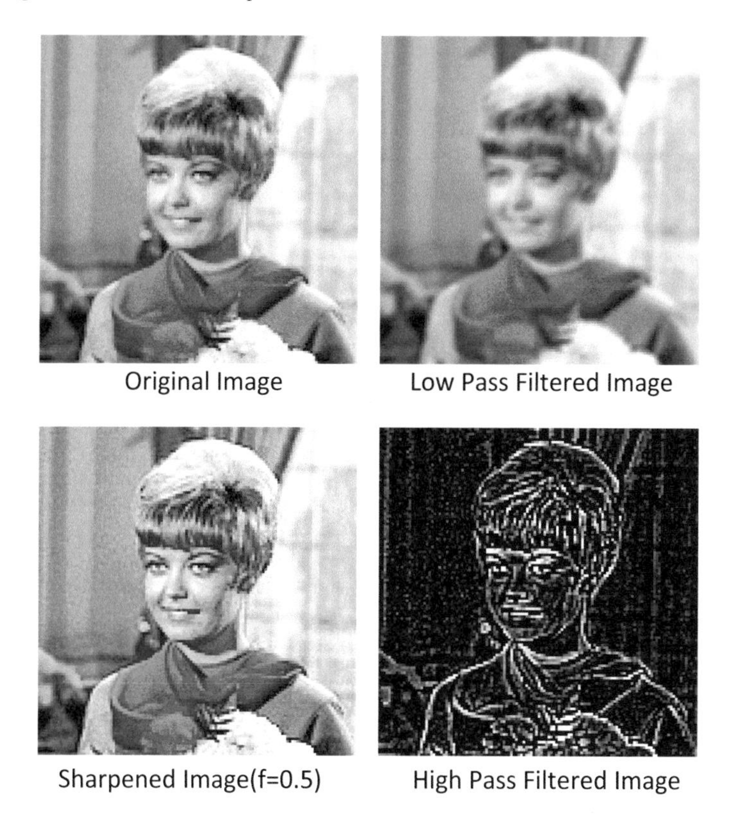

Fig. 11.1 Example illustrating the effect of digital image filtering (after Fred Weinhaus [2])

In general, the weights depend on i, j, resulting in a filter which varies across the image.

The above example is for windows with odd numbers of rows and columns. For even-sized windows, there will be a half-pixel displacement between the input and output images.

11.5 Digital Linear Filter in the Frequency Domain

Representation of digital linear filter in the frequency domain is also known as the Fourier representation. Here instead of representing an image as an $n \times n$ array of pixel values, we represent it as the sum of many sine waves of different frequencies, amplitudes, and directions. The parameters specify the sine waves are the Fourier coefficients. The reason of using this type of representation is that some linear filters can be computed more efficiently in the frequency domain by using the Fast Fourier Transform (FFT).

11.5.1 Theory of the Fourier Transform

The Fourier Transform of f, given by f* is an n × n array given by

$$f_{kl}^{*} = \sum_{i=1}^{n} \sum_{j=1}^{n} f_{ij} \exp\left\{-\frac{2\pi \varsigma}{n}(\mathrm{ik} + \mathrm{jl})\right\} \tag{11.9}$$

for k, l = 1, …, n and $\varsigma = \sqrt{-1}$.

The inverse Fourier Transform recovers f from f*:

$$f_{ij} = \frac{1}{n^2} \sum_{k=1}^{n} \sum_{l=1}^{n} f_{kl}^{*} \exp\left\{\frac{2\pi \varsigma}{n}(\mathrm{ik} + \mathrm{jl})\right\} \tag{11.10}$$

for i, j = 1, …, n.

The Fast Fourier Transform (FFT) is a very fast implementation of the Fourier Transform. The column transform is given by

$$f_{kj}^{\prime} = \sum_{i=1}^{n} f_{ij} \exp\left(\frac{2\pi \varsigma}{n}\right) \mathrm{ik} \tag{11.11}$$

for k = 1, …, n.

can be performed separately for each column j, from 1 to n, using a real-only implementation of the FFT.

The subsequent row transform will be

$$f_{kl}^{*} = \sum_{j=1}^{n} f_{kj}^{\prime} \exp\left\{-\frac{2\pi \varsigma}{n} \mathrm{jl}\right\} \tag{11.12}$$

for l = 1, 2, 3, …, n, which requires a complex FFT algorithm (11.8) but needs only be performed for k = n/2, …, n because of the structure of f*.

11.6 Nonlinear Digital Filter

Linear smoothing filters blur edges because both edges and noise are high frequency components of images. Nonlinear digital filters are potentially more powerful than linear filters because they are able to reduce noise levels without simultaneously blurring edges. However, they can generate spurious features and distort existing features in images and they can be computationally expensive. The simplest and the most widely used nonlinear filter is the moving median filter.

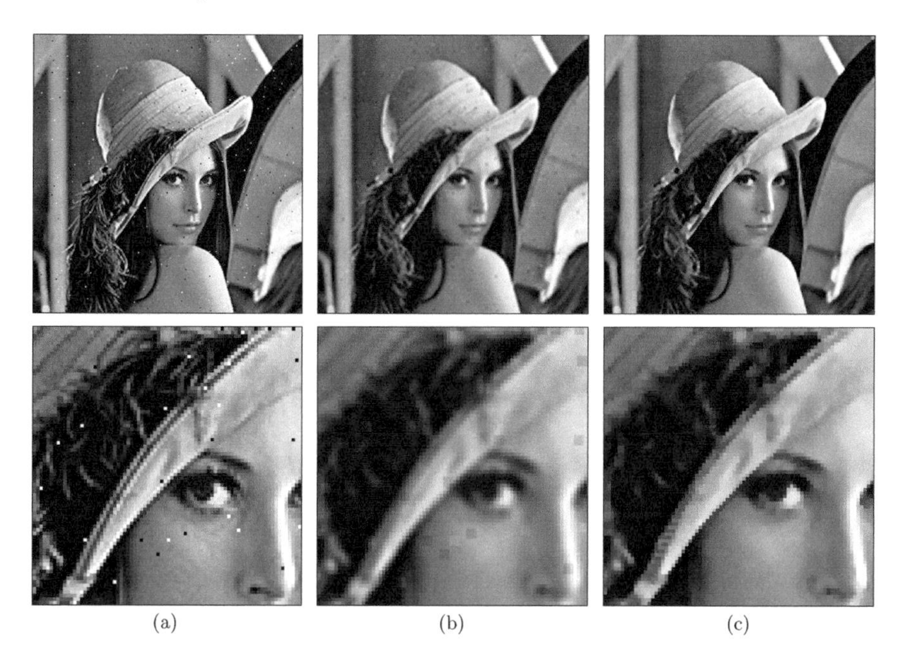

Fig. 11.2 Shows the effects of median filter in noise removal (after Agu [3])

The moving median filter is similar to the moving average filter, except that produces as output at a pixel the median, rather than the mean of the pixel values in a square window centred around that pixel. For a nonlinear digital filter of size (2m + 1) × (2m + 1), the output will be

$$g_{ij} = \text{median}\left\{f_{i+k,j+l} : \text{k}, \text{l} = -\text{m}, \ldots, \text{m}\right\} \quad (11.13)$$

Nonlinear filters are not additive. That is repeated application of a median filter is not equivalent to a simple application of a median filter using a different size of window. By making repeated use of a nonlinear smoothing filter with a small window, it is sometimes possible to improve on noise reduction while retaining fine details which would be lost if a larger window was used (Fig. 11.2).

References

1. Gonzadez, R. C., & Woods, R. E. (2002). *Digital image processing*. Boston: Addison-Wesley.
2. Lecture Notes by Fred Weinhaus.
3. Agu, E. Lecture 4: Filters (Part 2) & Edges and Contours. Digital Image Processing (CS/ECE 545). Worcester Polytechnic Institue, Computer Science Department.

Chapter 12
Digital Image Enhancement

12.1 The Purpose of Image Enhancement

Image enhancement [1] is a subfield of digital image processing. The purpose of image enhancement is to improve the contrast and sharpening the image to enable for further processing or analysis. It is the purpose of adjusting digital images so that the results are more suitable for display or further image analysis. For example, the removal of noise, sharpening or brightening an image, making it easier to identify key features. Image enhancement improves the quality and the information content of original data before processing. The enhancement does not raise the inbuilt information content of the data other than it increases the dynamic range of the selected facial appearance as a result that they can be detected.

Overall image enhancement techniques can be classified under two main categories:

A. Spatial Domain Enhancement Method

Here domain means the collection of pixels composing an image. Spatial domain techniques are procedures that work directly on these composed pixels.

B. Frequency Domain Method

This type of image enhancement has the frequency filters which developed the image in the frequency domain. Then the following steps are taken:

i. Transformation of the image into the Fourier domain.
ii. Multiply the image by the filter.
iii. Perform the inverse transform of the image.

W. S. Gan, *Signal Processing and Image Processing for Acoustical Imaging*,
https://doi.org/10.1007/978-981-10-5550-8_12

12.2 Image Segmentation

Image segmentation is the procedure of partitioning a digital image into multiple segments of sets of pixels. The purpose is to simplify and change the representation of an image into something that is easier and more meaningful to analyze. For instance it is used to locate objects and boundaries such as lines, curves etc. in images. It assigns a label to every pixel in an image such that pixels with the same label share certain characteristics.

The result of image segmentation is a collection of segments which collectively cover the entire image or a set of contours extracted from the image giving rise to edge detection. Each of the pixels in a region arse similar with respect to some characteristics such as texture, intensity, or colour. Adjacent regions are significantly different with respect to the same characteristics. An example is as in medical imaging, one can apply this image segmentation to a stack of images, resulting in contours that can be used for 3D image reconstruction with the use of interpolation algorithms like marching cubes.

An example of the application of image segmentation to acoustical imaging is medical ultrasound imaging. This will enable the location of tumours and other pathologies, the diagnosis, and study of anatomical structures, and the measurement of tissue volumes. The image segmentation can be classified into the two basic types: global segmentation, concerning with segmenting the whole image, consisting of large number of pixels and local segmentation, concerning with specific region or part of image. There are several general purpose algorithms and techniques developed for image segmentation.

12.2.1 Histogram-Based Method of Image Segmentation

In this method, a histogram is computed from all of the pixels in the image. The peaks and valleys in the histogram are used to locate the clusters in the image. This method is very efficient compared to other image segmentation methods because only one pass through the pixels is required. Intensity or colour can be used in the measurement.

The histogram-based method can be refined by recursively applying it to clusters in the image in order to divide them into smaller clusters. This is a common technique to improve the performance of large images by down sampling the image into clusters. Then compute the clusters and reassign the values to the larger image.

The procedure is repeated using smaller and smaller clusters until no more clusters are formed. This histogram-based method has the disadvantage that it may be difficult to identify significant valleys and peaks in the image.

An advantage of this histogram-based approach is that it can be quickly adapted to apply to multiple frames while maintaining their single pass efficiency. With multiple frames, the histogram can be done in multiple fashion. By applying to multiple frames, after the results are merged, one can identify peaks and valleys that were previously difficult to identify and are now more likely to be distinguishable. One can also apply this approach on a per-pixel basis with the resulting information to be used for the determination of the most frequent colour for the pixel location.

12.2.2 Edge Detection Method of Image Segmentation

Edge detection technique is another well established field in image segmentation. The boundaries and edges of various regions in the image are closely related due to the sharp adjustment in intensity at the region boundaries Closed region boundaries are required to segment an object from an image. The edges identified by edge detection are often disconnected.

The image segmentation methods can be categorized into two types based on properties of image: the discontinuity detection based approach and the similarity detection based approach. The discontinuity detection based approach segments the image into region based on discontinuity. An example of this is the edge detection method in which edges formed due to intensity discontinuity are detected and linked to form boundaries of region. The similarity detection based approach segments the image into regions based on similarity. An example is the clustering method which divides the image into set of clusters having similar set of pixels.

The edge based segmentation methods are based on the rapid change of intensity value in an image because a single intensity value does not provide good information about edges. This technique locates the edges based on either the first derivative of intensity is greater than a particular threshold or the second derivative has zero crossings. To apply this method, first the edges are detected and connected together to form the object boundaries to segment the required regions.

12.2.3 Clustering

The clustering based techniques are the techniques segmenting the image into clusters having pixels with similar characteristics. This is a form of data clustering which divides the data into clusters so that the elements in the same cluster will be similar to each other. Clustering methods can be divided into the hierarchical method and the partition based method: the concept of trees is introduced in the hierarchical method with the roots of the tree represents the whole database and the clusters are represented by the internal nodes. The partition based method, on the other hand, use optimisation methods iteratively to minimise the objective function.

There are basically two types of clustering:

1. Hard Clustering

This is a clustering technique that divides the image into sets of clusters with the condition that one pixel can only belong to only one cluster. This means that each pixel can belong to exactly one cluster. Membership functions are used with that values either 1 or 0 with 1 meaning certain pixel can belong to particular cluster and 0 means otherwise.

2. Soft Clustering

In real life application exact division is not possible due to the presence of noise. Soft clustering is a more natural type of clustering and hence is most useful for image segmentation in which division is not strict. An example is the fuzzy c-means clustering in which the pixels are partitioned into clusters based on partial membership or one pixel can belong to more than one clusters. This type of belonging is described by membership values. Hence this technique is more flexible than other techniques.

12.3 Contrast Enhancement

The quality of an image can be improved without introducing irrelevant artifacts. The global contrast enhancement method will increase the luminance for bright pixels and decreases the luminance for the dark pixels. Therefore the neighbourhood dependent contrast enhancement is desirable to get enough contrast for image enhancement without losing the dynamic range compression. As a whole, contrast enhancement can be classified under the following categories:

A. Linear Contrast Enhancement

Here the original image can be linearly expanded into a new distribution. By expanding the original value of an image, the total range of sensitivity of the digital device can be utilized. This enhancement method is also known as contrast stretching. It is mostly used in remote sensing images.

B. Nonlinear Contrast Enhancement

This type of enhancement method involves the histogram equalization method through the use of an algorithm. It has the limitation that each value of input image has several values in the output image and due to this the original object will loose their accurate brightness.

12.3.1 Histogram Equalization

The histogram equalization (HE) is one of the methods of contrast enhancement of an image. It can be divided into two main categories: the local histogram equalization and the global histogram equalization. For the local method the overall contrast of an image can be enhanced more effectively. During histogram equalization processing, the pixels are modified by transformation function based on gray level content of an entire image. The shape of the histogram changes throughout the process of histogram equalization. The general formula for the calculation of histogram equalization is given by

$$\mathrm{h}(\nu) = \mathrm{round}\left(\frac{cdf(\nu) - cdf_{min}}{(M \times N) - cdf_{min}} \times (\mathrm{L} - 1)\right) \tag{12.1}$$

where $\mathrm{cd}f_{\mathrm{min}}$ = minimum non-zero value of the cumulative distribution function, $\mathrm{M} \times \mathrm{N}$ = the image's number of pixels and L = number of gray level used. The advantage of histogram equalization is that it is straightforward, effective, and simple with an invertible operator. It is a popular image enhancement algorithm applied to medical imaging, geographical information system etc.

12.3.2 Adaptive Histogram Equalization

This is another method of image contrast enhancement. This is used to reduce the problem of the poor brightness of an image. The parameter of this technique is the size of the neighbouring region. For the adaptive histogram equalization (AHE), at smaller scales, the contrast of an image is enhanced, while at larger scales, the contrast of an image is reduced (Fig. 12.1).

The advantage of adaptive histogram equalization is that it is locally adaptive, automatic, and reducible.

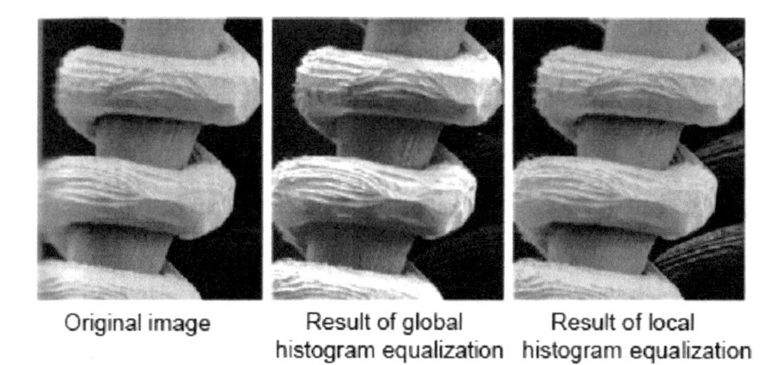

Fig. 12.1 The results of adaptive histogram equalization. **a** Original image, **b** output image (after [2])

12.3.3 Fuzzy Based Image Enhancement

Fuzzy based image enhancement is an important technique of digital image processing. Fuzzy logic is a form of artificial intelligence. It is based on fuzzy sets. It is used to the development of human awareness in the form of fuzzy if then rules. This method is one of the most effective and efficient methods for improving the image quality. Fuzzy methods can deal with the improbability and limitations of an image that can be represented as a fuzzy set. Fuzzy image processing can be divided into three sections: image fuzzification to encode images data, membership values modifications and image defuzzification to decode image data. The quality of poor images can be improved (Fig. 12.2).

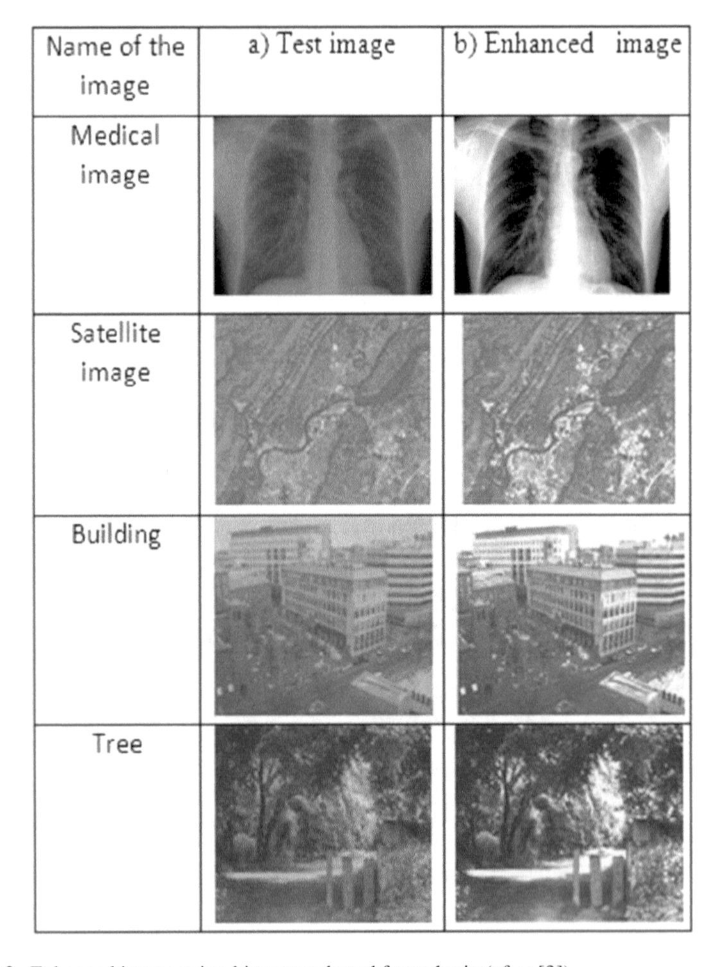

Fig. 12.2 Enhanced image using histogram based fuzzy logic (after [3])

References

1. Wikipedia.
2. Maini, R., & Aggarwal, H. (2010). A comprehensive review of image enhancement techniques. *Journal of Computing*, *2*(3), 8–13.
3. Sarath, K., & Sreejith, S. (2017). Image enhancement using fuzzy logic. *IOSR Journal Electronics and Communication Engineering,* 34–44.

Chapter 13
Deep Learning and Artificial Neural Networks

13.1 Introduction

Artificial neural networks is an important tool used in artificial intelligence and machine learning or deep learning. An example of their applications is to image recognition. They are brain-inspired and are intended to replicate the way humans learn. Neural networks consist of input and output layers as well as a hidden layer consisting of units that transform the input into a form that the output layer can use. They are suitable to find patterns that are far too complex or numerous for a human programmer to extract and teach the machine to recognize. The patterns recognized are numerical, contained in vectors into which all real world data, be it images, sound, text or time series must be translated. Neural networks have been around since the 1940s but only in the last several decades that they have been become a major part of artificial intelligence. This is due to the arrival of the new technique of backpropagation which enables one to adjust the hidden layers of neurons in the neural networks in situations when the outcome does not match what the created is expected.

Another step of advance is the arrival of deep learning neural networks in which it has the capacity that different layers of a multilayer network can extract different features until it is able to recognise what it is searching for (Fig. 13.1).

As an example, if one applies the deep learning neural network to a factory line, with the data set represent the raw materials as input, and then passed down the conveyor belt, with each subsequent layer extracting a different set of high-level features. If the neural network is to recognize an object on the factory line, then the first layer will analyze the brightness of its pixels. The next layer will then identify edges in the image, based on lines of similar pixels. Then the next layer will recognize textures and shapes etc. When arriving at the fourth or fifth layer then the deep learning neural network will be able to detect complex features of the image.

After this has been done, then the network has been trained, and various labels can be given to the output and backpropagation technique will be used to correct mistakes that have been made. After sometime, the network has been trained and

W. S. Gan, *Signal Processing and Image Processing for Acoustical Imaging*,
https://doi.org/10.1007/978-981-10-5550-8_13

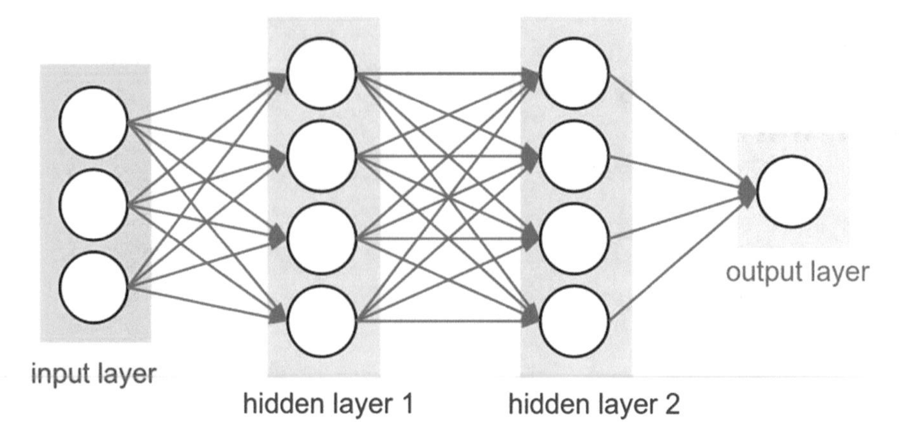

Fig. 13.1 Cross section of a multilayer neural network

able to carry out its own classification tasks without requiring the help of humans every time.

There are various types of neural networks for different specific uses and different levels of complexity. The most basic type is the feedforward neural network which is unidirectional with information travels from input to output. There is another type in which data can flow in multiple directions known as the recurrent neural network. This will enable greater learning abilities and are used for more complex tasks such as learning handwriting or language recognition. This type of neural network is more widely used. Besides these, there are other types such as the Hopfield networks and the convolutional neural networks. To choose the correct neural network, one depends on the specific application and the data one has to train it with. In some cases, multiple approaches have to be used such as dealing with voice recognition.

Neural networks are designed for identifying patterns in data such as classifying data sets into predefined classes, clustering data into different undefined categories and prediction of future events by using past events to govern future ones. Neural networks will need data to learn. The more data one provides for the network, the more accurate it will be. Also by repeating the process, it will become more efficient and make few mistakes.

To train a neural network, one has to divide the data into three sets. The first is a training set, which enables the network to establish the various weights between its nodes. Then the validation data set is used to fine-tune it. At the end, one will use a test set to check whether the network can successfully turn the input into the desired output.

One of the challenges in using neural network is the amount of time required to train the networks. This will require a considerable amount of computing power for more complex tasks. The biggest challenge for neural network can only fine-tune the answers when the user feeds in data and receives the answers. It does not have access to the exact decision making process.

Neural networks are a set of algorithms that are designed to recognise patterns. Sensing data are interpreted through a kind of machine perception, labeling or clustering raw input.

One can consider neural network as a clustering and classification layer on top of the data one stores and manages. They can help to group unlabelled data according to similarities among the example inputs and the data are classified when they have a labelled data set to train on. Neural networks are also able to extract features that are fed to other algorithms for clustering and classification. Thus one can consider deep neural networks as components of layer machine-learning application including algorithms for reinforcement learning, classification, and regression.

13.2 Training of Neural Network

A. Supervised Learning

This means that humans must transfer their knowledge to the data set in order for a neural network to learn the correlation between labels and data. Any labels that humans can generate any outcomes that one cares about and which correlate to data, can be used to train a neural network.

The applications can be to:

i. Identify objects in images such as the pedestrian stop signs, pedestrian lane markers etc.
ii. To recognise faces, facial expressions and to identify people in images.
iii. To detect voices, to identify speakers, and to transcribe speech to text.
iv. To recognise sentiment in voices.
v. To classify text as spam in emails or as fraudulent in insurance claim.
vi. To recognise sentiment in text as customer feedback.
vii. To recognise gestures in video.

B. Unsupervised Learning

Unsupervised learning is learning without labels clustering or grouping and is the detection of similarities. Deep learning does not require labels to detect similarities. The majority of data in the world are unlabelled. In machine learning, the more data an algorithm can train on, the more accurate it will be. Hence, unsupervised learning is able to produce highly accurate models.

Unsupervised learning can be used to:

i. Search: to compare images, sounds or documents to reveal similar items.
ii. Anomaly detection: anomalies are unusual behaviours. Anomalies can reveal fraud which one has to detect and prevent.

13.3 Predictive Analysis: Regression

Deep learning is also able to establish correlation between pixels in an image and the name of a person. This is known as static prediction. In the same manner, with sufficient data, deep learning is also able to establish correlation between present events and future events. It can run regression between the past and the future. Deep learning is independent of time. It can read a string of numbers from a time series and predict the number most likely to occur. Examples that it can predict are:

i. Health breakdown such as heart attacks or strokes based on vital statistics and data from measurable.
ii. Hardware breakdowns such as in data centres, in transportation, and in manufacturing.
iii. Employee turnover.
iv. Predicting the likelihood that a customer will leave based on web activity and metadata.

13.4 Elements of Neural Networks

Another name of deep learning is stacked neural networks. This means neural networks consist of several layers. These layers are made of nodes. A node is loosely patterned to neuron in the human brain. It fires when it encounters sufficient stimuli. It is the place where computation occurs. A node will combine the data which is input with a set of coefficients known as weights. The weights will either amplify or dampen that input. This will assign significance to the inputs with regard to the task the algorithm is trying to learn such as which input is most useful in classifying data without error. These input-weight products are then summed and passed through the node's activation function for classification to determine whether and to what extent that signal should progress further through the networks to affect the ultimate outcome. The neuron will be activated if the signal can pass through.

This procedure can be illustrated by Fig. 13.2.

A node layer is like a row of switches made of neurons that will turn on or off as the input is fed through the net. Starting from an initial input layer receiving the data, each layer's output is simultaneously the subsequent layer's input. One can assign significance to those features with regard to how the neural network classifies and clusters input by passing the model's adjustable weights with input features.

13.5 Deep Neural Networks

The commonplace neural network is single hidden-layered. Deep learning neural networks possess the depth concept represented by a multi-numbered node layers

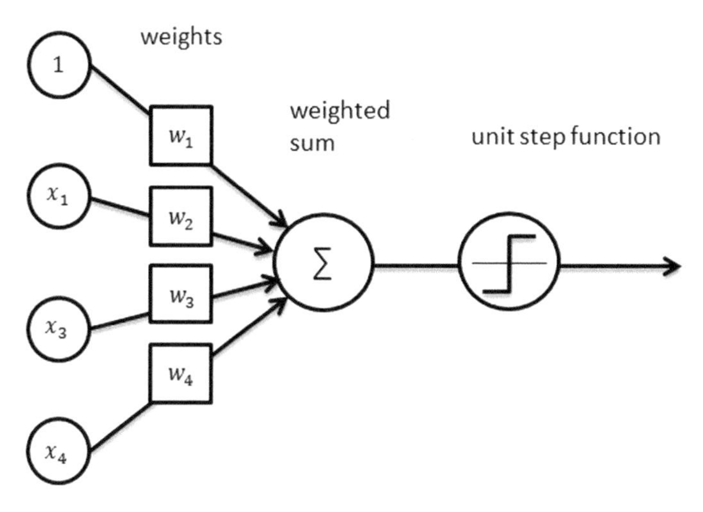

Fig. 13.2 Node from neural network

through which data will pass in a multistep process of pattern recognition. Early version of neural networks such as the first perceptrons, composed of only one input layer, one output layer, and at most one hidden layer in between. More than total three layers will be qualified as deep learning.

Each layer of nodes trained on a distinct set of features based on the previous layer's output. The further one advances into the neural net, the more complex the features the nodes can recognize, since they aggregate and recombine features from the previous layer. This is known as feature hierarchy (Fig. 13.3).

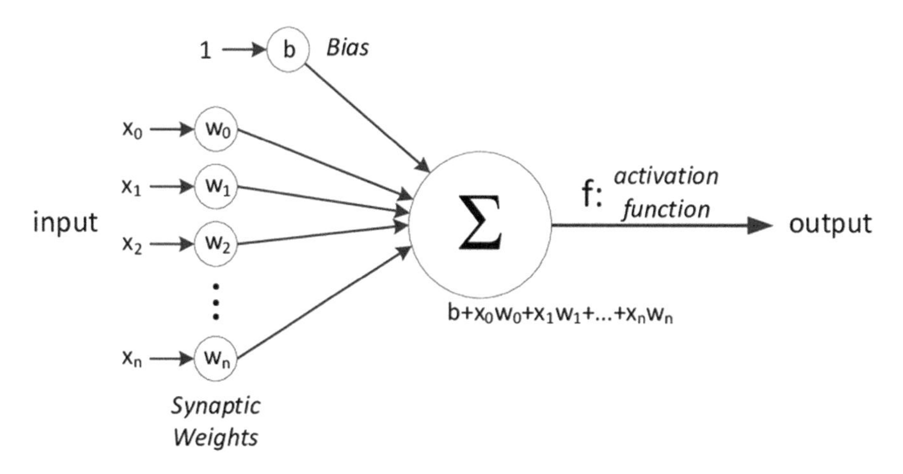

Fig. 13.3 Perceptron

This is a hierarchy of increasing complexity and abstraction. This will enable deep learning neural networks to handle very large, high-dimensional data sets with billions of parameters that pass through nonlinear functions.

Above all, these neural networks are able to discover latent structures, within unlabelled, unstructured data, which are raw data from pictures, audio recordings, texts, and video. Therefore deep learning will solve best in processing and clustering the raw, unlabelled data, discerning anomalies and similarities in data that no human has organized in a relational database. Deep learning can a take a million images and cluster them according to their similarities for instance the so-called smart photo album. The same idea can be applied to other data types such as clustering raw text such as news articles or emails forming the basis of various messaging filters, and can be used in customer-relationships management (CRM). For instance, emails from satisfied customers might cluster in one corner of the vector space, emails full of angry complaints or spam messages might cluster in others.

The clustering of data with time series can be applied to anomalous dangerous behaviour and normal/healthy behaviour. With the time series data generated by a smart phone, insight into users' health and habits will be provided. It might be used to prevent catastrophic breakdown if it is being generated by an autopart.

13.6 Automatic Feature Extraction

Automatic feature extraction is a capability of deep learning neural networks. Most traditional/without human interventions machine learning algorithms are not able to do this. Complex feature extractions that take teams of data scientists years to accomplish can be circumvented by deep learning. It will augment the power of data science teams which do not scale. During the training process on unlabelled data each node layer in a deep neural network will learn features automatically by repeatedly trying to reconstruct the input from which it draws its samples. It will attempt to minimize the difference between the network's guesses and the probability distribution of the input data itself. These neural networks learn to recognize correlations between certain relevant features and optimal results. They draw connections between feature signals and what these features represent, including cases with full reconstructions or with labelled data. After training the deep learning network on labelled data, it can then be applied to unstructured data. This will allow it access to much more input than machine learning nets. The more data a net can train on, the more accurate it will be. In fact, bad algorithms trained on lots of data ca outperform good algorithms trained on fewer data. Compared with previous algorithm, deep learning has the advantage of able to process and learn from huge quantities of unlabeled data. Deep learning neural networks will end in an output layer known as a logistic or softmax. This is a classifier that assigns a likelihood to a particular outcome or label. This is known as predictive. For example, if the raw data is an image, then deep learning network can decide that the input data is 90% likely to represent a person.

13.7 Feedforward Neural Networks

The goal of a neural net is to arrive at the point of least error as soon as possible. This is like running around a loop in a race and passing the same points repeatedly in a loop. The starting line for the race is the state in which the weights are initialized. The finish line is the state of those parameters when they are capable of producing sufficiently accurate classifications and predictions. Each step of the neural net involves an error measurement, and a slight update on its weights which is an incremental adjustment to the coefficients as it slowly learns to pay attention to the most important features. It is a repetitive act.

A neural network model is a collection of weights whether they are in their beginning or end state. The purpose is to model the data's relationship to ground truth labels to obtain the data's structure. As the neural network updates its parameters over time, the model will be improved. This because it does not know which weights and biases that will translate the input best to obtain the correct guesses. This is like starting with a guess and as it learns from its mistakes, better guesses will be sequentially achieved.

After input enters the neural networks, the weights or coefficients will map that input to a set of guesses the network makes at the end.

$$\text{Input} * \text{weight} = \text{guess} \tag{13.1}$$

Weighted input will result in a guess about what that input is. The neural network will take its guess and compare this to a ground-truth about the data, which is asking an expert the question: did I get this right?

$$\text{ground truth} - \text{guess} = \text{error} \tag{13.2}$$

The above error is given by the difference between the network's guess and the ground truth. The neural network will measure that error then walks the error back over its model and then adjusts weights to the extent that they contributed to the error.

$$\text{error weight's contribution to error} = \text{adjustment} \tag{13.3}$$

The three equations above describe the three key functions of the neural networks: obtaining input, calculating loss and applying update to the model. After this the three-step process is repeated all over again. Hence a neural network is a corrective feedback loop, that punishes weights that lead to error and rewards weights that support its correct guesses.

13.8 Multiple Linear Regression

Although artificial neural networks are biologically inspired, they are basically in terms of mathematics and codes, similar to other machine-learning algorithms. In fact, they work according to linear regression, a common method in statistics represented by

$$\text{Y\_hat} = \text{bX} + \text{a} \tag{13.4}$$

where Y_hat = estimated output, X = input, b = slope, and a = intercept of a line on the vertical axis of a two-dimensional graph. The linear relationship means that every time one adds a unit to X, the dependent variable. Y_hat increases proportionally, no matter how far along one is on the X axis. This can be extended to multiple linear regression where one has many input variables producing an output variable such as variable Y_hat increases proportionally, no matter how far along one is on the X axis. This can be extended to multiple linear regression where one has many input variables providing an output variable such as:

$$\text{Y\_hat} = \text{b\_1}^*\text{X\_1} + \text{6\_2}^*\text{X\_2} + \text{b\_3}^*\text{X\_3} + \text{a} \tag{13.5}$$

One can use the practical example of X presenting the amount of fertilizer, amount of sunlight, and rainfall, and Y_hat = size of crop. This form of multiple linear regression is happening at every node of a neural network. For each node of a single layer, input from each node of the previous layer is recombined with input from every other node. Therefore the inputs are combined in different proportions, according to their coefficients, leading into each node of the subsequent layer. This is the way the neural network tests which combination of input is significant in order to reduce error.

If every node merely performed multiple linear regression, this means Y_hat would increase linearly and without limit as X's increase. This does not fulfil our purpose. Hence one will need to build at each node, a switch such as a neuron that will turn on and off, depending on whether or not it should allow the signal of the input to pass through to achieve the ultimate decision of the network.

There is a classification problem with the switch. That is to classify the input's signal that indicates the node as on or off, enough or not enough. There is a binary decision that can be expressed by 1 and 0 and the logistic regression is a nonlinear function that squashes input to translate it to a space between 0 and 1. The nonlinear transformations at each node are usually s-shaped functions similar to logistic regression. Some of their names are sigmoid, tanh, hard tanh etc. The shape of each node depends on the s-shaped function.

The output of all nodes each squashed into an s-shaped space between 0 and 1, will be passed as input to the next layer in a feedforward neural network. This process will continue until the signal reaches the final layer of the net, where decisions are made.

13.9 Gradient Descent

Gradient means slope. It represents two variables are related to each other on an x–y graph. It describes how the error varies as the weight is adjusted or the relationship between the network's error and a single weight.

The purpose is to find out which weigh will produce the least error and correctly represents the signals contained in the input data, and translates them to a correct classification. In the process of learning, a neural network will slowly adjust many weights so that they can map signal to the correct meaning. The relationship between the network error and each of those weights is a derivative, dE/dW which measures the degree a slight change in a weight will cause a slight change in the error. Each weight is just one factor in a deep neural network. It involves many transformations so that the signal of the weight passes through many activations and sums over several layers.

The chain rule of calculus can be used to march back through the networks activations and outputs to arrive finally at the weight in question and its relationship to overall error.

The chain rule in calculus can be given by

$$\frac{dz}{dr} = \frac{dz}{dy} \cdot \frac{dy}{dr} \tag{13.6}$$

In a feedforward neural network, the relationship between the net's error and a single weight will be given by:

$$\frac{dError}{dweight} = \frac{dError}{dactivation} \cdot \frac{dactivation}{dweight} \tag{13.7}$$

Hence given two variables, Error and weight, that are mediated by a third variable, activation, through which the weight is passed, one can calculate how a change in weight will affect a change in Error by calculating how a change in Error and how a change in weight can affect a change in activation.

Deep learning is a learning process of adjusting a model's weights in response to the error it produces until one cannot reduce the error any more.

13.10 Logistic Regression

The final layer of a deep neural network with many layers, has a particular role. With labelled input, the output will classify each example and apply the most likely label. Each node on the output layer represents one label. That node will turn on or off according to the strength of the signal it receives from the previous layer's input and parameters.

Each output node will produce the binary output values 0 or 1 because an input variable either deserves a label or it does not.

Although neural networks working with labeled data produce binary output, the input is often continuous and will span a range of values including any number of metrics, depending on the problem to be solved.

The mechanism to be used to convert continuous signals into binary output is logistic regression. It calculates the probability that a set of inputs match the label.

13.11 History of Artificial Neural Networks

Artificial neural networks started in 1943 with McCulloch and Pitts [1] creating a computational model for neural networks. Then Hebb [2] in the late 1940s, created a learning hypothesis based on the mechanism of neural plasticity which became known as Hebbian learning. In 1954, Farley and Clark [3] used computational machines, called 'calculators,' to simulate a Hebbian network. In 1958, Rosenblatt [4] created the perceptron [4]. In 1965 the first functional network with many layers were published by Ivakhnenko and Lapa [5] as the group method of Data Handling. In 1960 Kelley [6] derived the basics of continuous backpropagation [6] in the context of control theory.

13.12 Applications of Artificial Neural Networks

Artificial neural networks are able to reproduce and model nonlinear processes. So they have found applications in many disciplines in areas including system identification and control, pattern recognition, face identification, signal classification, 3D reconstruction, object recognition, medical diagnosis and more. They have been used to diagnose cancers, including lung cancer [7], prostate cancer, colorectal cancer [8] and to distinguish highly invasive cancer cell lines from less invasive lines using only cell shape information [9]. Artificial neural networks have also been used for building black box models in geoscience, hydrology [10], ocean modelling and costal engineering [11], and geomorphology [12].

Artificial neural networks can be used to study short- term behaviour of individual neurons [13] such as how behaviour can arise from abstract neural modules that represent complete subsystems and the dynamics of neural circuitry arising from the interaction between individual neurons. It can also be used as a tool to simulate the properties of many-body open quasi-systems [14].

References

1. McCulloch, W., & Walter, P. (1943). A logical calculus of ideas immanent in nervous activity. *Bulletin of Mathematical Biophysics*, *5*(4), 115–133.
2. Hebb, D. (1949). *The organization of behaviour*. New York: Wiley.
3. Farley, B. G., & Clark, W. A. (1954) Simulation of self-organizing systems by digital computer. *IRE Transactions on Information Theory, 4*(4), 76–84.
4. Rosenblatt, F. (1958). The perceptron: A probabilistic model for information storage and organization in the brain. *Psychological Review*, *65*(6), 386–408.
5. Ivakhnenko, A. G., & Lapa, V. G. (1967). *Cybernetics and forecasting techniques*. New York: American Elsevier Pub. Co.
6. Kelley, H. J. (1960). Gradient theory of optimal flight paths. *ARS Journal*, *30*(10), 947–954.
7. Ganesan, N. (2010) Application of neural networks in diagnosing cancer disease using demographic data. *International Journal of Computer Applications*, *1*(26), 81–97.
8. Bottaci, L. (1997). Artificial neural networks applied to outcome prediction for colorectal cancer patients in separate institutions. *The Lancet*, *350*, 469–472.
9. Alizadeh, E., Lyons, S. M., Castle, J. M., & Prasad, A. (2016). Measuring systematic changes in invasive cancer cell shape using Zernike moments. *Integrative Biology*, *8*(11), 1183–1193.
10. ASCE Task Committee on Application of Artificial Neural Networks in Hydrology. (2000). Artificial neural networks in hydrology. I: Preliminary concepts. *Journal of Hydrologic Engineering, 5*(2), 115–123.
11. Dwarakish, G. S., Rakshith, S., & Natesan, U. (2013). Review on applications of neural network in coastal engineering. *Artificial Intelligent Systems and Machine Learning, 5*(7), 324–331.
12. Ermini, L., Catani, F., & Casagli, N. (2005). Artificial neural networks applied to landslide susceptibility assessment. *Geomorphology, 66*(1), 327–343.
13. Forrest, M. D. (2015). Simulation of alcohol action upon a detailed Purkinje neuron model and a simpler surrogate model that runs >400 times faster. *BMC Neuroscience, 16*(27), 27.
14. Yoshioka, N., & Hamazaki, R. (2019). Constructing neural stationary states for open quantum many-body systems. *Physical Review B, 99*(21), 214306.

Chapter 14
Practical Cases of Applications of Artificial Intelligence to Acoustical Imaging

The applications of artificial intelligence (AI) or deep learning will revolutionize modern healthcare system. It is the application of artificial neural networks to medical ultrasound imaging. AI can also have industrial applications such as the diagnosis of images in nondestructive testing. In modern healthcare, the two most important elements are AI assisted screening and AI-assisted consultation. First an AI healthcare platform is needed. Patients will describe their symptoms on the platform. Besides collecting the information, the AI system also asks further questions that could be relevant for possible treatments. This will be a pre-diagnosis to forecast ailments and a preliminary report will be issued by the AI system. This report will be sent to a doctor who conducts a professional diagnosis and decides on further steps. The doctor may describe medicine or send the patient to a specialist in a clinic. This is a seamless transition and the patient may not even notice the switch between the actual doctor and the AI system. Thus the decision made by the doctor together with the consultation protocol are saved and entered into a machine learning system. Thus the AI system can further improve on its diagnosis and prediction with the goal of being able to treat patients completely independently of human input. There will be a significant reduction in waiting time for offline services.

AI can be used to improve diagnostic accuracy and efficiency. It could be used more in intensive care units where it could help alert hospital staff to possible incidents quickly by detecting anomalies in patients' vital signs. This can help lower mortality rates. It can also increase the speed of image detection. For instance Google has developed a system for grading prostate cancer that does it more accurately than US pathologists and a Stanford team has achieved similar success with skin cancer. Where lots of data exist and precision is valued, AI can help humans make better decision, even though it still messes up regularly when trained on biased data set or is intentionally tricked. Humans are less prone to misidentifying objects and are better able to correct for their biases.

An example of the applications of AI to ultrasound imaging for diagnostics is as follow. A team from Germany, the United States and France [1] taught an AI system to distinguish dangerous skin lesions from benign ones showing it in more than 100,000 images. There are about 232,000 new cases of melanoma and 55,500 deaths

W. S. Gan, *Signal Processing and Image Processing for Acoustical Imaging*,
https://doi.org/10.1007/978-981-10-5550-8_14

in the world each year [2]. Melanoma incident rates are rising steadily in fair-skinned population and were predicted to further increase [2]. It showed that a computer was better than human dermatologists at detecting skin cancer. The ultrasound imaging systems using a deep learning convolutional neural network or CNN was tested against 58 dermatologists from 17 countries showed photos of malignant melanomas and benign moles. A deep learning CNN was trained, validated and tested for the diagnostic classification of dermoscopic images of lesions of melanocytic origin (melanoma, benign nevi). The diagnostic performance of the dermatologists will be improved when they received more clinical information and images.

Most dermatologists were outperformed by the CNN. On average, dermatologists accurately detected 86.6% of skin cancers from the images compared with 95% for the CNN. The CNN missed fewer melanomas meaning it had a higher sensitivity than the dermatologists. It also misdiagnosed fewer benign moles as malignant melanoma. This would result in less unnecessary surgery. The dermatologists' performance improved when they were given more information of the patients and their skin lesions. Hence AI showed to be able to provide faster and easier diagnosis of skin cancer, allowing surgical removal before it spreads.

However, it is unlikely that the AI machine can take over from human doctors entirely. It can also function as an aid. For instance, melanoma in some parts of the body such as the fingers, toes and scalp are difficult to image and AI will have difficulty recognizing "atypical" lesions or ones that patients themselves are unaware of.

A reproducible high diagnostic accuracy would be desirable for early melanoma detections. The German team trained and tested the convolutional deep learning CNN for differentiating dermoscopic images of melanoma and benign levi. The diagnostic performance of the CNN was compared with a large international group of 58 dermatologists from 17 countries and 30 of the experts had more than 5 years of experience. At study level-I when the dermatologists were provided with only dermoscopic images, their dichotomous classification of lesions was significantly outperformed by the CNN. In real-life situation, the dermatologists will be provided more clinical information into the decision-making. Hence a study-level II is performed when additional clinical information and close-up images were given. It showed a much improved diagnostic performance of the dermatologists. However, even at their improved mean sensitivity of 88.99%, the dermatologists still showed a specificity inferior to the CNN. This study showed that CNN algorithm is a useful tool to aid physicians in melanoma detection irrespective of their individual level of experience and training. It demonstrates that an adequately trained deep learning CNN is capable of a highly accurate diagnostic classification of dermoscopic images of melanocytic origin. Although a CNN's architecture is difficult to set up and train, its implementation on digital dermoscopy system of smart phone application can easily be deployed. Hence physicians of different levels of experience and training can be benefited from the CNN's image classification.

Another example of the application of AI to ultrasound imaging system is for the location of position to inject antiseptics for cesarean delivery. This will help the injection of anesthesia up to 92%. Currently, before the injection of anesthesia the doctor will use hand to touch on the spine to find the correct location. This is very demanding on the doctor's knowledge of the structure of the spine. This is especially difficult for the case of fat, abnormal and those undergone surgery before. With the use of AI the anesthesian only needs to use ultrasound imaging system to scan the whole spine from bottom to top, and the AI system will locate the correct position and give out the signal. This will highly improve the success of first time injection. Those women participants are with normal, low risk spines.

Usually for normal patients for anesthesia injection into spine, the success rate is only between 50 and 60% for first time injection. Using AI spine anesthesia system, the clinical test success rate is rather high. The AI system only reduces the fear, worries and unsuitability of the patients to injection. This AI system plays an important role in the training of anesthesia specialist in determining the spine injection position more accurately.

The side effects of several injections are spine blood swollen, skin with needle injection feeling, headache from spine membrane piercing through, and body not feeling well.

References

1. Haenssle, H. A., Fink, C., Schneiderbaue, R., Toberer, F., Buhl, T., Blum, A., et al. (2018). Man against machine: Diagnostic performance of a deep learning convolutional neural network for dermoscopic melanoma recognition in comparison to 58 dermatologists. *Annals of Oncology, 29*(8), 1836–1842.
2. Nicolaon, V., & Stratigos, A. J. (2014). Emerging trends in the epidemiology of melanoma. *British Journal of Dermatology, 170*(1), 11–19.

Printed in the United States
By Bookmasters